EARTH SCIENCES IN THE 21ST CENTURY

RIVER CHANNELS

TYPES, DYNAMICS AND CHANGES

EARTH SCIENCES IN THE 21ST CENTURY

Additional books in this series can be found on Nova's website
under the Series tab.

Additional E-books in this series can be found on Nova's website
under the E-book tab.

EARTH SCIENCES IN THE 21ST CENTURY

RIVER CHANNELS

TYPES, DYNAMICS AND CHANGES

DANIEL A. MOLINA

EDITOR

Nova Science Publishers, Inc.

New York

For permission to use material from this book please contact us:
Telephone 631-231-7269; Fax 631-231-8175
Web Site: http://www.novapublishers.com

Additional color graphics may be available in the e-book version of this book.

Library of Congress Cataloging-in-Publication Data

River channels : types, dynamics and changes / editor, Daniel A. Molina.
p. cm.
Includes index.
ISBN 978-1-61324-148-6 (hardcover)
1. River channels. 2. Geomorphology. I. Molina, Daniel A.
GB561.R594 2011
551.48'3--dc22

2011009087

Published by Nova Science Publishers, Inc. † New York

CONTENTS

PREFACE

This book presents current research in the study of the types, dynamics and changes of river channels. Topics discussed include zoobenthos diversity in the Volga and Danube European rivers; large alluvial river channel patterns and complexity; channel evolution, fluvial management and safety maintenance in Russian river channels; alluvial river channel morphology in New Zealand and river planform analysis as a tool for neotectonic investigation in the northern Pannonian Basin.

Chapter 1 - The Volga, with a length of 3531 km and a basin area of 1,431 Million km^2, is the largest river in Europe and the most important waterway in Russia. The Volga basin is the nucleus of the Russian civilisation and comprises nowadays 40 % of the population, 45 % of the country's industry and 50 % of its agriculture. Especially the 20th century brought a significant change for the Volga River: a cascade of hydroelectric power plants was constructed along its course and therefore most of the hydrobiological studies in the Volga basin were focusing on lentic water bodies, i.e. reservoirs. However, recently the free-flowing section in the headwaters between the Upper Volga Lakes and the city of Tver was identified as a refuge area for the potamal fauna of European importance.

The Danube is about 2850 km long and thus it is the second largest river in Europe. It arises in the Black Forest in Germany and discharges into the Black Sea. According to the International Commission for the Protection of the Danube Basin (ICPDR) the catchment with a total area of 801,463 km^2 includes the territories of 19 countries and is the home of 83 million people with a wide range of cultures, languages and historical backgrounds. The ecosystems of the Danube River Basin are highly valuable in environmental, economic, historical and social terms, but they are subject to increasing pressure and serious pollution from agriculture, industry and cities. The above mentioned numbers illustrate and emphasise the enormous international importance of this river.

In this chapter the main interest focuses on the investigation of biodiversity of macroinvertebrates of the free-flowing sections, i.e. the number of species and their relative proportions in the whole invertebrate community, as well as major governing environmental factors. A study of biodiversity is of fundamental importance within the field of ecology, in order to address questions such as why some species are numerous and others are rare, and what kind of influence their relative abundances have on the structure and function of the whole community. Furthermore a review on research history, longitudinal specifics, faunistical changes and invasion pathways as well as management guidelines will be presented.

Chapter 2 - The geometry of an alluvial river channel is controlled by 1) the complexity of the hierarchy of alluvial forms (the number of levels of hierarchy); 2) the height of alluvial forms relative to the channel depth; 3) the width of alluvial forms relative to the channel width; 4) the hierarchic level of stable (vegetated) alluvial forms; 5) river basin climate and vegetation; wash load concentration in the stream; 6) river valley geological structure and texture; 7) the history of the river valley evolution. Interaction between the river flow and channel bed is the main cause of alluvial forms generation. The hierarchy of alluvial forms consists of 3-8 levels. The complexity of the system generally increases with the river width/depth ratio increase and slope decrease. The typical sequence of alluvial forms is: 1) megaripples (ultramicroforms); 2) dunes of the first and second order (microforms); 3) large dunes of the third order and small bars of the first order (mesoforms); 4) large bars (macroforms). These alluvial forms (usually meso- and macroforms), depending on their relative height, river basin climate and vegetation, and wash load concentration in the stream, can be stabilised by the vegetation and determine the channel pattern. River channel can be called 'meandering' when the width (lateral wavelength) of the largest stabilised (vegetated) alluvial feature, which determines overall channel geometry, is close to the channel double width. Meanders are simple when only these largest bars of hierarchical complex are vegetated. During the low water period smaller non-vegetated bars appear above the water level and the river can be characterised as meandering with alternating and braid bars. The river channel can be called braided by islands if the width (lateral wavelength) of the largest stabilised (vegetated) alluvial feature, which determines channel pattern, is less than the channel double width. The larger the relative channel width is and the lower the hierarchical level of the vegetated alluvial form is, the more complicated the braiding is. During the low water the channel can be braided both by islands and non-vegetated bars. The river channel is braided by bars (at low water) or straight (at high water) when all alluvial forms are non-vegetated and overall channel geometry is controlled only by non-fluvial factors or/and old fluvial relief. Non-fluvial control factors (soil texture and rocks, vegetation in the river valley and slopes), as well as old fluvial relief of floodplain and river terraces, increase the complexity of the river channel patterns.

Chapter 3 - Though various classifications have been proposed to impose some rational sort of classification on varieties of fluvial processes, only a few could be regarded as universal ones. The list of classification systems, developed by significant contributors before the early 2000s for both stream and valley types is not small; the given ones represent a new tool in hydromorphological research. It was derived using data from large numbers of rivers from various environments of the former USSR and thus applied both to high-gradient mountain, debris-flow, rocky, plain, and marshy rivers. The classification is obtained by descriptions of the channel forms according to their spatial scale. Fluvial processes and channel patterns types are used to describe channel level of the fluvial classification. Five main fluvial process types in the stream corridor (from high-gradient streams with rapid/falls to plain rivers) and 3 groups of channel patterns (straight, meandering, and braided) are found. All of the types are separated using a variety of quantitative geomorphologic indicators (stream order, stream power, stream size and channel slopes, entrenchment ratio, and sinuosity or amount of branches).

The above-mentioned classifications are much different in the given channel type's quantitative and qualitative descriptions. Variety of views on channel types determines that uncertainties exist in fluvial system management between different researchers and practitioners. The underlying problems is to find the unique approach to the variety of channel forms, to derive the complete analytical and empirical solution for the sediment entrainment, transport and deposition and to develop the specific assessment of stream stability, channel evolution, and potential response for each channel type. The goal of this paper is to estimate the state-of-the-art methods in fluvial classification and describe the mostly full approach to detail the variety of rivers and channels.

The final conclusion that river channels should be described through comprehensive qualitative and quantitative descriptions of the all its controls, forms, and constituents. The complex approach enables researchers to find places for the object and thus classify it or put it in an order and further apply this information for channel evolution, fluvial management, and safety maintenance. The corresponding examples are given in the paper.

Chapter 4 - River management in New Zealand's laterally active gravelly rivers has permitted floodplain development and protection of agricultural resources and infrastructure. Management of these dynamic systems has been hailed as a success for the approaches adopted, namely straightening and confining the river using bank protection. However, this activity also impacts river morphological / habitat diversity and potential gravel resource, by replacing broad riparian corridors with narrower channels and reduced lateral connectivity with the floodplain. This chapter quantifies river behaviour at a range of spatial and temporal scales by assessing geomorphic change in the lower Rangitikei River over 50 years and in the upper Motueka River over a 5 year period, using decadal aerial photo analysis and annual RTK-dGPS surveys respectively to address the effect of river management on channel morphology.

The lower Rangitikei has been transformed from a braided to predominantly single-thread wandering planform. This narrowed channel has remained relatively stable since 1983. Similarly, the upper Motueka has also been narrowed since 1960. Flood management practices such as river training by placement of willows and rock protection have stabilised these river reaches.

To quantify geomorphic change and associated sediment transfers more precisely in these narrowed channels, annual RTK-dGPS surveys between 2004 - 2010 acquired data to construct Digital Elevation Models (DEMs) in a ~1 km reach of the upper Motueka River. Morphological budgeting based on subtraction of successive DEM surfaces reveals complex spatial and temporal patterns of erosion and deposition. Overall, volumetric changes suggest these narrowed reaches have been net exporters of sediment, associated with continued channel degradation. This has left bar features, traditionally the focus of gravel extraction in the reaches, relatively isolated from all but extreme flows, limiting replenishment of the gravel resource.

Chapter 5 - Sinuosity calculations were made in the Upper Theiss region to draw conclusions on the neotectonic activity of the northern-north-eastern part of the Pannonian Basin.

The upstream part of the River Theiss, the Bodrog and its tributaries (Ondava, Laborc, Ung and Latorca), the Sajó and the Hernád flows in the study are. These rivers were meandering before the river controls. Their last, pre-regulated planforms were mapped in the second part of the 19th century, during the Second Military Survey of the Habsburg Empire.

The sinuosity values were computed along these rivers with different sample section size (window size). This window size changed from 5 km to 80 km. There is a given window size for each rivers, determined by the bank-full discharge, which connected to the neotectonics. Along these rivers, the characteristic window size was 15 or 20 km.

The places, where the river sinuosity changed, were compared to the structural lines of the "Atlas of the present-day geodynamics of the Pannonian Basin". Every place where the fault lines of the neotectonical map crossed the rivers, the sinuosities were changed, increased or decreased depending of the direction of the relative vertical movements.

Chapter 6 - The Nile is the longest river on earth, it flows for nearly 6700 km from Latitude 4° S(the Equatorial Lakes) to Lat 32° north (the Mediterranean Sea) and it also drains the largest catchment in Africa, which equates an area of 31,1 million km^{2}. Proxy data of the fluvial sediments, cores of the Equatorial and Ethiopian lakes, age dating techniques, indicate that the river system has undergone great changes in recent geological history. The regional tectonic movements and geological evolution of the area have controlled the integration of different constituting catchments into the Nile mega drainage basin. Moreover, the Quaternary climatic changes have manifested drastic hydrological fluctuations, variability in the Nile flows and connectivity of different catchments via overflow. The Equatorial and Rift African lakes were desiccated during arid climatic phases, but extended several tens or even hundreds of metres above their present levels, and thus enlarged contributing catchment areas to the Nile system during more prominent Monsoon and wet phases. Because the downstream riparian of the Nile Basin (Egypt and Sudan) are mostly dependent on the Nile River, it is necessary to assess the impact of hydro-geological setting, climatic variability and water management strategies of the headwater regions in the Nile Basin. In this paper remote sensing data, digital elevation models (DEMs) were utilised to extract the catchment parameters, and to delineate storage areas for the Equatorial Lakes region. Available rainfall/discharge data are integrated with the delineated morphometrical parameters of the catchments to construct a hydrological model, which simulates the water balance of different interconnect basins. The total annual discharges of the basins are generally very low (less than 7 % of the total annual rainfall). Although, the basin of the shallow Lake Kioga (5 m deep) is of negative hydrological contribution among the Equatorial Lakes region, where all the annual rainfall and considerable amount of bypass water overflown from Lake Victoria basin are usually lost by evaporation and evapotranspiration. The runoff coefficient and discharge can be significantly increased by draining the Kioga Lake and swamps through dredging and deepening certain overflow channel reaches. On the other hand, construction or towering of the regulation and hydropower dams on the equatorial lakes will be of adverse impact on the annual water discharge downstream. This includes the occasional reduction of flow required for filling up to designed storage capacities, and permanently increasing surface areas of water that is exposed to evaporation. Alternative sites across given stream rapids and falls were suggested as suitable for considerable dam-hydropower generations and minimal water storage.

In: River Channels: Types, Dynamics and Changes
Editor: Daniel A. Molina, pp. 1-50
ISBN 978-1-61324-148-6

Chapter 1

Zoobenthos Diversity and Recent Changes in the Major European Rivers: The Volga and the Danube

M. Schletterer [1,2]***, U. H. Humpesch*** [3] ***and W. Graf*** [4]
[1]TIWAG – Tiroler Wasserkraft AG, Division Engineering Services,
Eduard-Wallnöfer-Platz 2, 6020 Innsbruck, Austria
[2]University of Innsbruck, Institute of Ecology, Dpt. River Ecology and Invertebrate Biology, Technikerstraße 25, 6020 Innsbruck, Austria
[3]University of Vienna, Department of Freshwater Ecology,
Althanstrasse 141090 Vienna, Austria
[4]BOKU – University of Natural Resources and Life Sciences, Department of Water, Atmosphere & Environment, Institute of Hydrobiology & Water management, Working Group on Benthic Ecology and Ecological Status Assessment,
Max Emanuel-Strasse 17, 1180 Vienna, Austria

Abstract

The Volga, with a length of 3531 km and a basin area of 1,431 Million km², is the largest river in Europe and the most important waterway in Russia. The Volga basin is the nucleus of the Russian civilisation and comprises nowadays 40 % of the population, 45 % of the country's industry and 50 % of its agriculture. Especially the 20th century brought a significant change for the Volga River: a cascade of hydroelectric power plants was constructed along its course and therefore most of the hydrobiological studies in the Volga basin were focusing on lentic water bodies, i.e. reservoirs. However, recently the free-flowing section in the headwaters between the Upper Volga Lakes and the city of Tver was identified as a refuge area for the potamal fauna of European importance.

The Danube is about 2850 km long and thus it is the second largest river in Europe. It arises in the Black Forest in Germany and discharges into the Black Sea. According to the International Commission for the Protection of the Danube Basin (ICPDR) the catchment with a total area of 801,463 km² includes the territories of 19 countries and is the home of 83 million people with a wide range of cultures, languages and historical backgrounds. The ecosystems of the Danube River Basin are highly valuable in environmental, economic, historical and social terms, but they are subject to increasing pressure and serious pollution from agriculture, industry and cities. The above mentioned numbers illustrate and emphasise the enormous international importance of this river.

In this chapter the main interest focuses on the investigation of biodiversity of macroinvertebrates of the free-flowing sections, i.e. the number of species and their relative proportions in the whole invertebrate community, as well as major governing environmental factors. A study of biodiversity is of fundamental importance within the field of ecology, in order to address questions such as why some species are numerous and others are rare, and what kind of influence their relative abundances have on the structure and function of the whole community. Furthermore a review on research history, longitudinal specifics, faunistical changes and invasion pathways as well as management guidelines will be presented.

1. Introduction

Large river systems are important habitats and pathways for animals, links to adjacent floodplains and dynamic elements of the landscape (Ward et al. 2002, Dolédec & Statzner 2008). In contrast to their high value, little is known about their natural characteristics in Europe. But nowadays most large river-systems are threatened due to pollution, canalisation and changes in the hydrological regime, which even raised the question if a restoration of big rivers is possible (Gore & Shields 1995). Many rivers are affected by hydromorphological changes, such as the loss of connectivity between the main river and its floodplains, which together are limiting the survival, distribution and development of many species. As a consequence, the fauna of most Central European streams is degraded: generalists are predominant, while specialised species declined in their abundance – limnophilic species due to eutrophication and rheophilic ones, because they can not reach their lotic reproductive habitats (Fittkau & Reiss 1983, Aarts et al., 2003). The largest European Rivers - Volga, Danube and Rhine - serve as migration pathways for animals. Nowadays due to man-made channels, three main routes for the invasion of ponto-caspian fauna elements to Europe exist: (1) the northern corridor: Volga - Neva - Baltic Sea, (2) the central corridor connecting the rivers Dnieper - Vistula - Oder - Elbe - Rhine, and (3) the southern corridor connecting the Danube and Rhine (Gruszka 1999, de Vaate et al. 2002, Panov et al. 2005). Especially due to international transport via navigation new colonisation routes for non native faunas were opened, which presents another strong impact on the already weakened invertebrate community. Aquatic species belong to the most endangered organisms on an European scale due to manifold and complex habitat degradation tangling the factors organic and toxic pollution, straightening, damming and other hydromorphological impacts, as well as population pressures by invasive species. Rates of habitat modification are currently so high that virtually all natural habitats and protected areas are destined to become ecological 'islands' in surrounding 'oceans' of altered habitats. This process of fragmentation and isolation in landscapes under human influence – main concepts in island biogeographic theory – is predicted to lead directly and indirectly to accelerated species extinctions at both the local and the global scales, thus reducing the world's biodiversity at all levels (Lawton & May 1995). In the context of the so called 'McDonaldization' of the biosphere (Lövei 1997) the dispersal of many species is inhibited, while others – mostly more flexible species in ecological terms - become common and overtake the niches of indigenous species. Regarding aquatic ecosystems similar processes have already been observed by Fittkau & Reiss (1983), Zwick (1984, 1992) and Fochetti & Tierno de Figueroa (2006, 2008).

Most of the theories and concepts related to invertebrates, in the context of the ecology of running waters, are based on studies on small streams, whereas investigations of large rivers have played a minor role for a long time, mainly due to methodological difficulties. However, nowadays large rivers came into the focous, e.g. Statzner et al. (1994) provided a test of habitat templet theories for the Upper Rhone River.

This chapter provides a review on research history of the Volga and Danube and data about the zoobenthos communities, considering biodiversity, governing environmental factors, longitudinal specifics and recent changes. The general view on the Volga River is based on Behning`s (1924, 1928) historical descriptions, who provided a dataset for the Volga River between Tver and its mouth into the Caspian Sea, before the construction of the Volga cascade. Further, data from Butorin & Mordukhai-Boltovskoi (1978) as well as a number of recent studies provide a view on longitudinal changes. The present review about the free-flowing section in the headwaters of the Volga is considering studies by Schletterer & Eggers 2006, Schletterer et al. 2007, Zhenikov et al. 2007, Schletterer 2009, Schletterer & Füreder 2009, 2010a, 2010b, Schletterer et al. 2010). Also for the Danube historical campaigns are summarized. The basis of the review on the Austrian Danube are the studies by Fesl (2002) Fesl & Humpesch (2003, 2006a, b), Humpesch & Fesl (2005), Fesl et al. (1999) and Fesl et al. (2005). The present detailed review on macroinvertebrates in the free-flowing section of the Danube below Vienna provides an excellent opportunity to survey or restate scientific hypotheses on the basis of a large river – complementing the knowledge about the Austrian Danube, because the results of studies about the impounded areas of this river have been previously summarised by Humpesch (1992). Summarizing characteristics of the whole Danube are mainly based on the JDS2 report (Graf et al. 2008a).

2. Characteristics of the Volga Basin and Research History

The Volga River has a length of 3531 km, a basin area of 1,431 Million km^2 (Figure 1) and a mean annual discharge of 8346 m^3 s^{-1} (MQ at Volgograd 1879-1958, according to Shiklomanov 1999), which characterises it as largest river in Europe and the 16th largest river in the world. According to a common classification the Volga River was divided in (a) the Upper Volga from the spring down to the mouth of R. Oka, (b) the Middle Volga between R. Oka and R. Kama and (c) the Lower Volga down to its mouth into the Caspian Sea (BSE 1951). Nowadays the Gorky dam is considered as the border between the upper and the middle course and the Kuibyshev dam between the middle and the lower Volga (Litvinov et al. 2009).

The Volga arises between St. Petersburg and Moscow in the Valdai hills. From the source at Volgoverkhovje (228 m) it runs through two small lakes (Small and Big Verkhit) and 10.5 km downstream the source it joins Lake Sterzh. This lake is the uppermost of four river lakes (Stersh – 18 km^2, Vselug – 30 km^2, Peno – 17 km^2, Volgo – 61 km^2). In 1843 at the outlet of the lowermost lake a dam was build to rise the waterlevel (Verkhnevolzhskoye reservoir), in order to regulate the flow in the Upper Volga for nautic purposes, however these lakes are nowadays considered as semi-natural (Schletterer & Füreder 2010b). Downstream this dam, the

Verknevolzhsky Bejshlot, the free-flowing section starts and stretches down to the city of Tver. Sometimes River Selizharovka, the outlet from Lake Seliger, is described as "second source of the Volga" (Shaporenko et al. 2002). Below Tver the "Volga cascade" extends downstream until Volgograd: Ivankovo Reservoir, Uglich Reservoir, Rybinsk Reservoir, Gorky Reservoir, Cheboksary Reservoir, Kuybyshev Reservoir, Saratov Reservoir and Volgograd Reservoir (for details see Khaiter 1995). The main tributaries of the Volga River are: Selizharovka, Vazuza, Tvertsa, Sosha, Dubna, Medveditsa, Nerl, Kashinka, Mologa, Sheksna, Kotorosl, Kostroma, Unzha, Oka, Kerzhenets, Sura, Vetluga, Sviyaga, Kazanka, Kama and Samara (see Table 1).

Figure 1. Basin of the river Volga (from: www.cabri-volga.ru, with permission).

Upstream Volgograd the distributary Akhtuba River splits off the Volga. With the construction of the Hydroelectric power plant Volgograd this situation was altered by the dam, though since that time an artificial channel was made to ensure water intake into Akhtuba River (Gorelov, pers. comm.). The Akhtuba River is about 537 km long with an average discharge of 153 m³/s. Between Akhtuba and Volga, the famous Volga- Akhtuba floodplain extends.

About 60 km downstream of Astrakhan the Volga River spreads into the Caspian Sea. This is the largest river delta in Europe, covering 27,224 km² and spreading across 160 km. Due to changing sea level, there are three distinct zones developed in the Delta: (1) the "Baer`s mounds" (sand dunes), (2) active and abandoned water channels and (3) broad platform extending up to 60 km offshore (Schletterer & Humpesch, in prep.). The four main channels flowing through the delta are the Bachtemir, the Volga, the Tambola and the Akthuba.

Table 1. Main tributaries of Volga River

Tributary	Mouth	Catchment	Length	MQ	Comments
Selizharovka	in Selizharovo	2,950 km²	36 km	20 m³/s	Left tributary
Vazuza	in Zubtsov	7,120 km²	162 km	35.9 m³/s	Right tributary
Tvertsa	in Tver	6,510 km²	188 km	55 m³/s	Left tributary
Shosha	near Konakovo	3,080 km²	163 km	8.5 m³/s	Right tributary
Dubna	in Dubna	5,350 km²	167 km	11 m³/s	Right tributary
Medveditsa	near Kimry	5,570 km²	259 km	41 m³ /s	Left tributary
Mologa	near Vesyegonsk	29,700 km²	456 km	124 m³/s	Left tributary
Sheksna	near Cherepovets	19,000 km²	139 km	172 m³/s	Left tributary
Kotorosl	in Yaroslavl	5,500 km²	132 km	30 m³/s	Right tributary
Kostroma	in Kostroma	16,000 km²	354 km	85 m³/s	Left tributary
Unzha	near Yuryevets	28,900 km²	426 km	158 m³/s	Left tributary
Oka	in Nizhni Novgorod	246,000 km²	1,480 km	1240 m³/s	Right tributary
Kerzhenets	near Lyskovo	6,140 km²	290 km	19.6 m³/s	Left tributary
Sura	in Vasilsursk	67,500 km²	841 km	260 m³/s	Right tributary
Vetluga	in Kozmodemyansk	39,000 km²	889 km	255 m³/s	Left tributary
Sviyaga	near Kazan	16,700 km²	375 km	1590 m³/s	Right tributary
Kazanka	in Kazan	2,600 km²	142 km	299 m³/s	Left tributary
Kama	near Kazan	517,000 km²	1,805 km	3750 m³/s	Left tributary
Samara	in Samara	46,500 km²	594 km	45.6 m³/s	Left tributary

Modern scientific research begun about 200 years ago (Pallas 1771), and the Volga River was object of several expeditions and research activities (e.g. Baer 1855, Kessler 1877). Sabaneev (1880) and Milashevich (1881), who studied freshwater molluscs of the Middle Volga respectively Oka, brought invertebrates into the scientific focus. Rossinsky (1892) studied the Moskva River and provided a first detailed investigation on a benthic invertebrate community in this region region. The first hydrobiological river research station in Europe, was founded 1900 in Saratov ("Biological Volga Station") and it became soon an important, international research center (Table 2). Expeditions from the Upper Volga to the Caspian Sea were organized, and the results are published in two books with an overview about the whole Volga River (Behning 1924, 1928).

The 20th century brought a change for the Volga River: a cascade of hydroelectric power plants was constructed along the course of the river (Figure 2) and though hydrobiological research was focusing on lentic water bodies, i.e. reservoirs (see Butorin & Mordukhai-Boltovskoi 1978). A couple of research institutions evolved in the Volga basin: The "State Research Institute of Lake and River Fisheries" (GosNIORH: the main institute in St. Petersburg has branches in Pskov, Novgorod, Vologda, Konakovo, Nizhny Novgorod, Kazan, Perm, Saratov and Volgograd), the "Institute of Biology of Inland Waters" (Borok), the "Institute of the Ecology of the Volga Basin" (Tolyatti) and "Tver Floating Biological Station", which was located over 30 years upstream Tver and was transferred in 2007 to the Rybinsk reservoir (descriptions of these research institutes can be found in Schletterer & Füreder 2011).

Table 2. The list of researchers, which visited the Hydrobiological Volga Station (Behning 1924) respectively worked there, shows that many important limnologists (e.g. August Thienemann) where among them

Algae	K.I. Meier, L.V. Reingard, N.N. Voronikhin
Phanarogamae	D.E. Janishevski, E.V. Beljakov
Protozoa	S.A. Ilovaisky
Bryozoa	Annandale, E. Marcus
Hydra	R. Kirkpatrick, V.M. Izaev
Spongillidae	R. Kirkpatrick, P.D. Rezvoi
Hypania	L.A. Senkevich
Ostracoda	H.A. Girshman, W. Klie
Harpacticidae	V.M. Rylov
Hydrachnidae	S. Thor
Araneae	W. Kulczynski
Nematoda	H. Micoletzky, L. Jägerskiöld
Oligochaeta	W. Michaelsen
Hirudinaea	G.G. Shegolev
Ephemeroptera	G. Ulmer, Fr. Klapalek, A.V. Martynov
Plecoptera	A.V. Martynov
Trichoptera	G. Ulmer, A.V. Martynov
Odonata	E. Rousseau, A.N. Bartenev
Coleoptera	F.A. Saizev, H.L. Sacharow, H. Kuntzen
Mollusca	V.A. Lindholm
Chironomidae	A. Thienemann, J.J. Kieffer, F. Lenz
Simuliidae	G. Enderlein

At the edge of the 21st century Europe's largest river became the subject of interest for several international research projects: starting in 1992 with the "Oka-Elbe" project (Rupp et al. 2000), international collaboration was continued in 2005 with the research project "Volga-Rhine" (Middelkoop et al. 2005). The "Volga Vision" concentrated on socio-economic situation in the Volga Basin in 2000 and suggested targets to be reached within 30 years (Unesco 2004). Between 2004 and 2007 the CABRI (Cooperation Along a Big River) Volga project was run to develop action plans in cooperation with political stakeholders (Wefering & Nikitina 2005). A research expedition in 2005, focusing on the headwaters (Kuzovlev & Schletterer 2006), formed the basis for monitoring activities by Russian and Austrian scientists. And recently a project was carried out by the NWO (Netherlands Organisation for Scientific Research) in cooperation with Russian project partners in the large and almost intact Volga-Akhtuba floodplain: key-mechanisms for the annual fish reproduction related to year-to-year variation in the flood pulse were investigated (Gorski et al. 2010).

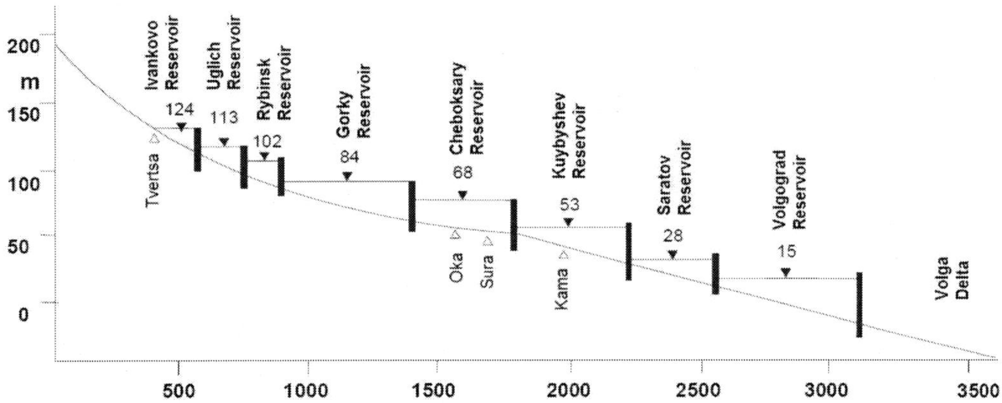

Figure 2. The reservoirs along the course of the Volga River (from Schletterer & Füreder 2011).

3. Characteristics of the Danube Basin and Research History

The Danube, as second largest river in Europe, is ca. 2850 km in length and according to the International Commission for the Protection of the Danube Basin (ICPDR 2009) its catchment covers an area of 801,463 km^2. Based on its gradients the Danube River Basin can be divided into three sub-regions: the upper basin, the middle basin, and the lower basin (including the Danube Delta) (Figure 3). A detailed description of the Danube basin was recently provided by Sommerfeld et al. (2009). The following overview about the Danube River, as well as its delta, is based on descriptions of the ICPDR (www.icpdr.com).

The *Upper Basin* extends from the source of the Danube in Germany to Bratislava in Slovakia and the Upper Danube River has a rapid current of between 8 and 9 km/hour, due to the pronounced gradient of the river bed. Its tributaries bring flow from the northern side of the Alps as well as out of the southern side of the Central European Highlands.

The *Middle Basin* extends from Bratislava to the dams of the Iron Gate Gorge at the border between Romania and Serbia and comprises the largest of the three sub-regions. In contrast to the upper reaches, the middle Danube looks like a flatbed river – low banks and up to 1.5 km wide, with around half the speed of the Upper Danube. Only in two stretches - at Visegrad (Hungary) and at the Iron Gates (at the boarder of Romania and Serbia) - the river flows through narrow, canyon-like gorges. After emerging from the Hungarian Gates Gorge near Bratislava, the Danube enters the Little Alföld Plain. In the plain the current slows down and the Danube loses its transporting capacity, leading to the deposition of great quantities of gravel and sand on the riverbed. On this alluvial fan two large islands developed, one on the Slovakian side (Žitný ostrov) of the river and the other on the Hungarian side (Szigetköz island), which have a combined area of about 1,869 km^2 and are home to 190,000 inhabitants in more than 100 settlements. Downstream Budapest the Danube flows across the vast Great Alföld Plain, with low terraces along both banks, until it reaches the Iron Gate Gorge. In this stretch there are many islands, including Csepel Island near Budapest, and three major tributaries (Drava, Tisza and Sava) join the Danube River, which nearly triples the flow of the mainstream.

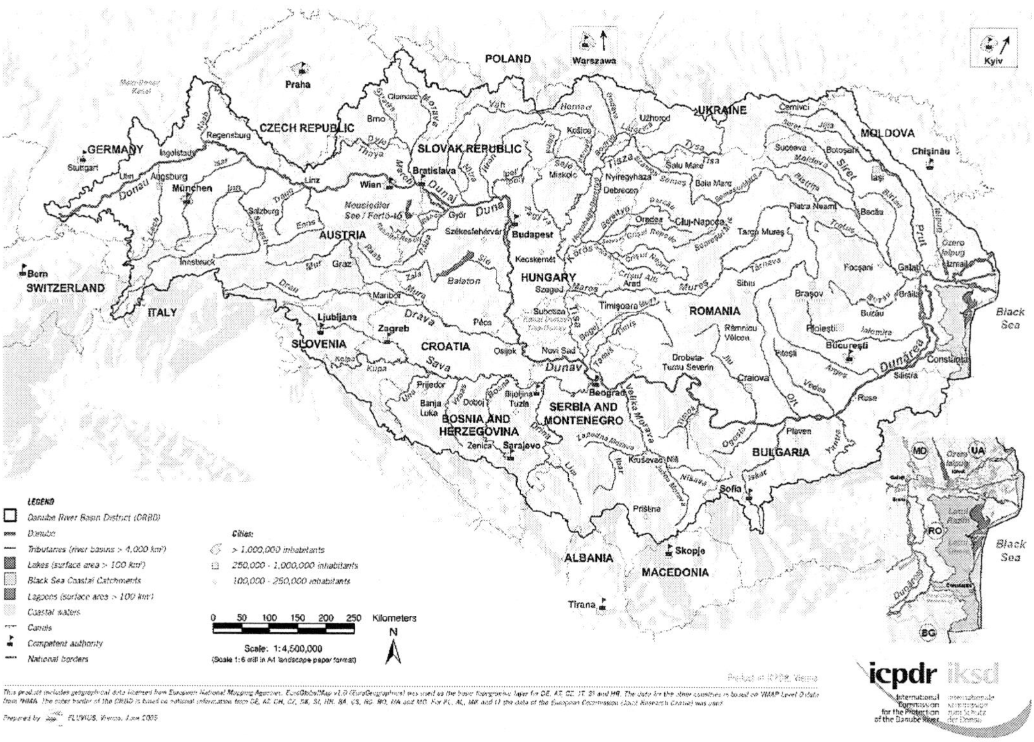

Figure 3. Danube River Basin: overview [with permission of the International Commission for the Protection of the Danube River (ICPDR)].

The *Lower Basin* of the River Danube includes plateaus, mountains and lowlands of Romania and Bulgaria. Beyond the Iron Gates, the Danube flows across a wide plain and becomes shallower and broader, with several large islands. Tributaries in this reach (including Iskar, Olt, Yantra, Siret and Prut) are comparatively small and account for only a modest increase in the total flow rate. Near Tulcea in Romania, some 80 km from the Black Sea, the river divides into three main branches – forming the *Danube Delta*, which covers an area of about 6,750 km². The three main channels flowing through the delta are the Chilia (carries 63 % of the total flow), the Sfintu Gheorghe (St. George; accounts for 21 % of the total flow), as well as the canalised and dredged Sulina channel (accounts for 16 % of the total flow), which is the only navigable one.

The main tributaries of the Danube River are Inn, Morava (March), Drava (Drau), Tisza (Theiss), Sava (Save), Iskar, Siret and Prut (in descending order according to the locations of confluences): The *Inn* is the third largest of the Danube's tributaries by discharge, and the seventh longest. At the confluence in Passau, it contains more water than the mainstream of the Danube itself, but its catchment area is with 26,130 km² less than half as big as the catchment of the Danube River at this point. The catchment of River *Morava* is lying between the Bohemian Highlands, the Carpathians and the Pannonian Plain. It extends over 26,658 km² and includes parts of the Czech Republic, Slovakia and Austria. The *Drava* River with a catchment area of 40,490 km², it is the fourth largest and with its length of 719 km fourth longest Danube tributary. It flows through Austria, Slovenia and Croatia, where it forms the border between Croatia and Hungary, before heading back into Croatia again to meet the Danube near Osijek. The Drava has been considerably regulated with dams and channels dredged to direct its flow. Nevertheless, natural habitats along the middle and lower reaches host unique assemblages of flora and fauna, including several endemic species – thus the whole Hungarian section of the river is protected as part of the Danube-Drava National Park. The *Tisza* River Basin, with its total extent of 157,186 km², is the largest sub-basin in the Danube River Basin. The Tisza River is with its 966 km the longest tributary of the Danube and with a mean annual discharge of 792 m³/s the second largest by flow. The *Sava* has the largest discharge of water (mean annual discharge = 1572 m³/s) to the Danube and with its basin of a size of 95,419 km², is it the second largest after the Tisza basin and with a length of 945 km the third largest tributary. The Sava is shared by Slovenia, Croatia, Bosnia and Herzegovina, Serbia and Montenegro. The *Iskar* has a total length of 368 km and a catchment area of 8,684 km², thus it is the largest Bulgarian tributary of the Danube. Its headwaters are located up in the passes of the Rila Mountain, further the mainstream flows through the suburbs of Sofia, and through the Balkan Mountains. The *Siret* River Basin with a catchment area of 44,835 km² is third largest catchment area of any Danube tributary, and is situated to the east of the Carpathians. The Siret River flows from the Ukraine to Romanis and has a length of 706 km. The source of *Prut* River is located in the forests of the Ukrainian Carpathians and later the river forms the border between Romania and Moldova. With its 959 km Prut River is the second longest but with a catchment of 27,500 km² and the last major tributary of the Danube, with its confluence located just upstream of the Danube Delta. From data given it can be concluded that the main tributaries of the Danube are quite different in hydrological characters.

According to Liepold (1961) research of the Danube started – at least in Austria – mainly with studies of the backwaters of the river (e.g. Steuer 1901, Brezina 1906, Schiller 1926, Pesta 1928, Vornatscher 1938, Oberzill 1941, Grohs 1943). Only a few ecological and biological studies were carried out by Brunnthaler (1900), Schallgruber (1943) and Stundl (1943, 1944a,b, 1951). See also “Die Pflanzen- und Tierwelt der Donau” by Starmühlner et al. (1972).

The international interest on the research of the Danube began when Liepolt founded the International Association for Danube Research in 1956 (www.iad.gs). In this association all countries linked to the Danube started to collaborate in a mainly qualitative and applied scientific frame-work. The outcome of this collaboration was published in a Supplement Volume of the Archiv für Hydrobiologie, called "Donauforschung". The name of the

Supplement was later changed into “Large Rivers”, through a further change the name of the journal is now “River Systems”. A summary of the first years of this international collaboration was presented in the monograph "Limnologie der Donau" (Liepolt 1965, 1966).

In 1998 an International Commission for the Protection of the Danube River was established to promote and coordinate sustainable and equitable water management practices, including conservation, improvement and rational water use (www.icpdr.org).

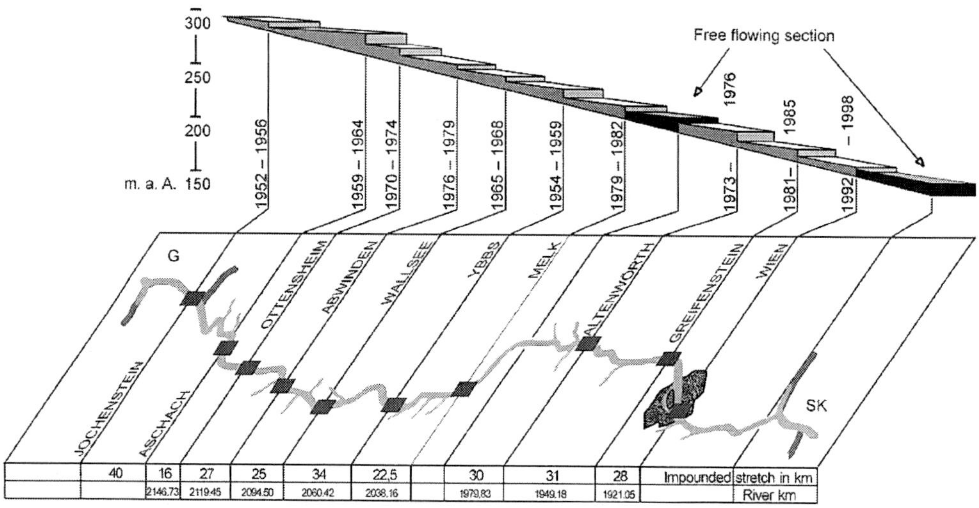

Figure 4. Hydroelectric power schemes of the River Danube in Austria, with dates of construction periods of power-stations (■) and impounded stretches: m.a.A. – metres above Adrian sea level (from Humpesch & Fesl 2005).

In Austria, research on the Danube was mainly stimulated when Liepolt founded the Austrian Committee of the International Association for Danube Research in 1976 (www.oen-iad.org). The Austrian part of the Danube descends in altitude by 156 metres over its 351 km length and, since the early 1950s, the river has been developed into a power-generating waterway (Figure 4), so that the continuity of the river is now interrupted by ten transverse barrages or dams used for hydroelectric purposes. As there are now ten impounded areas, only two stretches of the free-flowing river are left, the Wachau region (above river-km 2005), and the region downstream from the impoundment of Vienna (at river-km 1921). Therefore the Austrian stretch of the Danube is divided into three distinct macrohabitats: the free-flowing main river channel, back-waters of the impoundments, and backwaters of the riverside forests.

To study the effects of impounded stretches of the river, the Austrian Committee Danube Research initiated a study "to identify and to assess impacts of the hydropower scheme on the environment as well as an the regional infrastructure including social and economic aspects". This multidisciplinary study, "Ecosystem Study Altenwörth", was sponsored by the Austrian National Committee of the UNESCO-Man and Biosphere Programme (Grosina 1985, Hary & Nachtnebel 1989, Humpesch 1992, 1996, Parz-Gollner & Herzig 2000). Further stimulus to research on the Danube in Austria, mainly in respect to the free-flowing stretch of the river east of Vienna, came from the so-called "Committee on the Ecology of the Danube", which was established by the Austrian government as a result of controversy - started by the Green Movement in 1984 - around the planned hydropower scheme east of Vienna, in the riverside forests of Hainburg. Several studies were undertaken then, mainly contract work. Those which are linked more to the main channel and its riverside forests are: Interdisciplinary Study Danube (Hary & Janauer 1988); Ecotone Danube-March (Janauer 1996, Janauer & Hary 2003) and Monitoring Rivers and Floodplains (Schiemer & Janauer 1994); Strategies for Conservation of a Danube Fish Fauna (Schiemer & Waidbacher 1992); The importance of Riparian Ecotones for Diversity and Productivity of Riverine Fish Communities (Schiemer & Zalewski 1992); and Land/Inland Water Ecotones: Intermediate Habitats Critical for Conservation (Schiemer et al. 1995).

On an international scale ICPDR launched two large expeditions (JDS (Joint Danube Survey) 1 and 2) in the years 2001 and 2007 respectively which were the most homogenous chemical and biological analyses of the entire Danube from Germany to the Black sea along the main channel and selected tributaries ever carried out. While benthic invertebrates were sampled mainly by grabs in 2001 (Literáthy et al. 2002), air-lift and kick and sweep samples were taken in 2007 (Graf et al. 2008a). These expeditions provided an uniform dataset for the whole length of the river.

4. Longitudinal Changes of the Invertebrate Fauna along River Volga and River Danube

4.1. Volga

At the beginning of the last century data on the benthic communities of the Volga River was provided by Behning (1924, 1928), presenting a dataset of the Volga River before the construction of the hydropower cascade (Figure 5). Behning (1928) described different benthic communities:

- *Psammophil fauna:* This type was described as most common one (90 % of the river bed = sandy), extending from the headwaters down to the delta. Characteristic species were *Limnodrilus newaensis, Nais behningi, Gammarus sarsii, Gammarus platycheir, Cryptochironomus, Tanypus, Ancylus fluviatilis* and further downstream also *Enoploides fluviatilis* and *Dreissena polymorpha.*
- *Lithophil fauna:* This type was restricted to upper and middle reaches, where boulders from glacial times were deposited. In dense stands of the water moss *Fontinals antipyretica* in the Upper Volga species like *Ancylus fluviatilis, Asellus aquaticus, Aphelocheirus aestivalis, Ephemerella, Caenis, Brachycercus* and different Chironomidae were characteristic. Further downstream typically *Viviparus duboisianus, Dreissena polymorpha, Piscicola geometra, Corophium, Heptagenia sulphurea* and mainly Orthocladiinae dominated the faunal composition.
- *Pelophil fauna:* Muddy areas were dominated by *Dorylaimus stagnalis*, *Propappus volki* and *Limnodrilus newaensis, Erpobdella octoculata* and some chironomids. In the delta additionally "relicts of the Caspian Sea" occurred in this habitat type: *Neritina liturata, Hypania invalida, Jaera nordmanni* and Cumacaea.
- *Fauna associated with clay:* This type was found in the middle Volga and it was mainly dominated by burrowing mayflies *Ephoron virgo, Palingenia longicauda* and partly also *Ephemera vulgata.* Further *Hydropsyche, Corophium curvispinum,*

Dikerogammarus haemobaphes, Gammarus abbreviates, Viviparus and *Dreissena* occurred regularly.

- *Fauna associated with wood*: *Hydropsyche ornatula, Neureclipsis bimaculata, Polypedilum, Eukiefferiella, Dreissena polymorpha, Oligoneuriella rhenana, Ecdyonurus* and *Corophium* were found commonly associated with wood.
- *Biocoenosis of branches and distributaries:* Below the mouth of Kama a high number branches and side channels were developed down to the delta: these were dominated by *Viviparus duboisianus, Sphaerium rivicola, Sphaerium solidum, Limnodrilus newaensis, Limnodrilus hoffmeisteri, Glossiphonia complanata,* Mysidae, *Astacus leptodactylus, Aphelocheirus aestivalis, Gomphus flavipes, Chironomus thummi, Chironomus plumosus, Tanypus* and *Stictochironomus.*
- *Biocoenosis in the bays along the lower Volga:* Behning (1928) considered also the backwaters, which are only connected to the main river during high discharge. In these reaches *Unio, Anodonta, Sphaerium, Dorylaimus stagnalis, Limnodrilus hoffmeisteri* and *Erpobdella* were the characteristic benthic species.

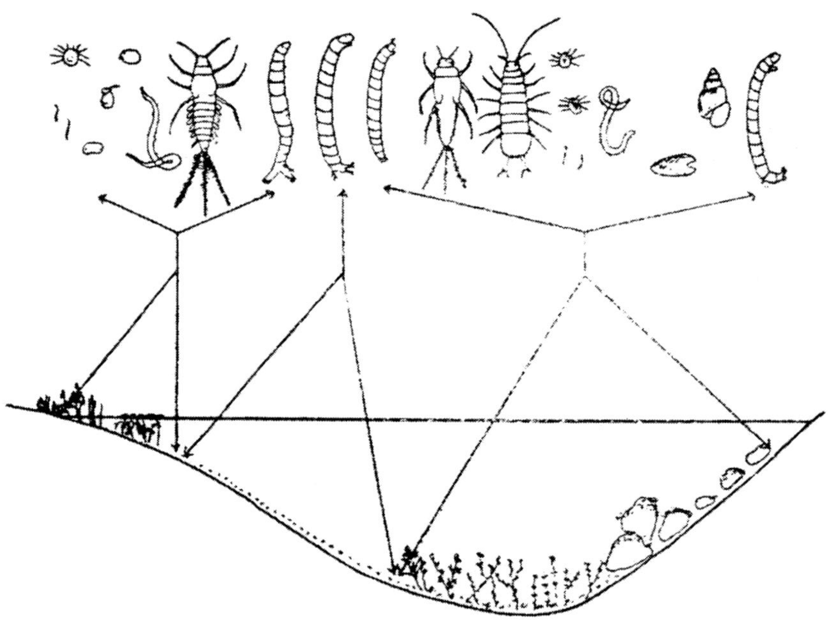

Figure 5. Schematic characterization of the Upper Volga's fauna (from Behning 1928).

The construction of the hydropower cascade along the Volga River brought a change to the riverine biota: characteristics of the Volga reservoirs were described by Butorin & Mordukhai-Boltovskoi (1978; see also references therein) and recently synthesized by Litvinov et al. (2009). Herein comparative data is provided for the free-flowing section in the headwaters and also an overview on the zoobenthos communities along the whole river is provided.

The free-flowing section between the Upper Volga Lakes and Tver remained intact, and hosts a diverse potamal fauna (see 5.1). The dataset assessed in August 2005 allows a comparison with the upper reach from the Danube (dataset provided by the JDS 2). Considering the number of species per taxa-group the dataset from 2005 shows a diverse faunal composition (Figure 6 a), however it has to be considered that – as some groups were not determined (e.g. Chironomidae, Oligochaeta) – the number of species will increase. This is shown e.g. for the monitoring sites (Figure 6 b), where 29 chironomid species were determined.

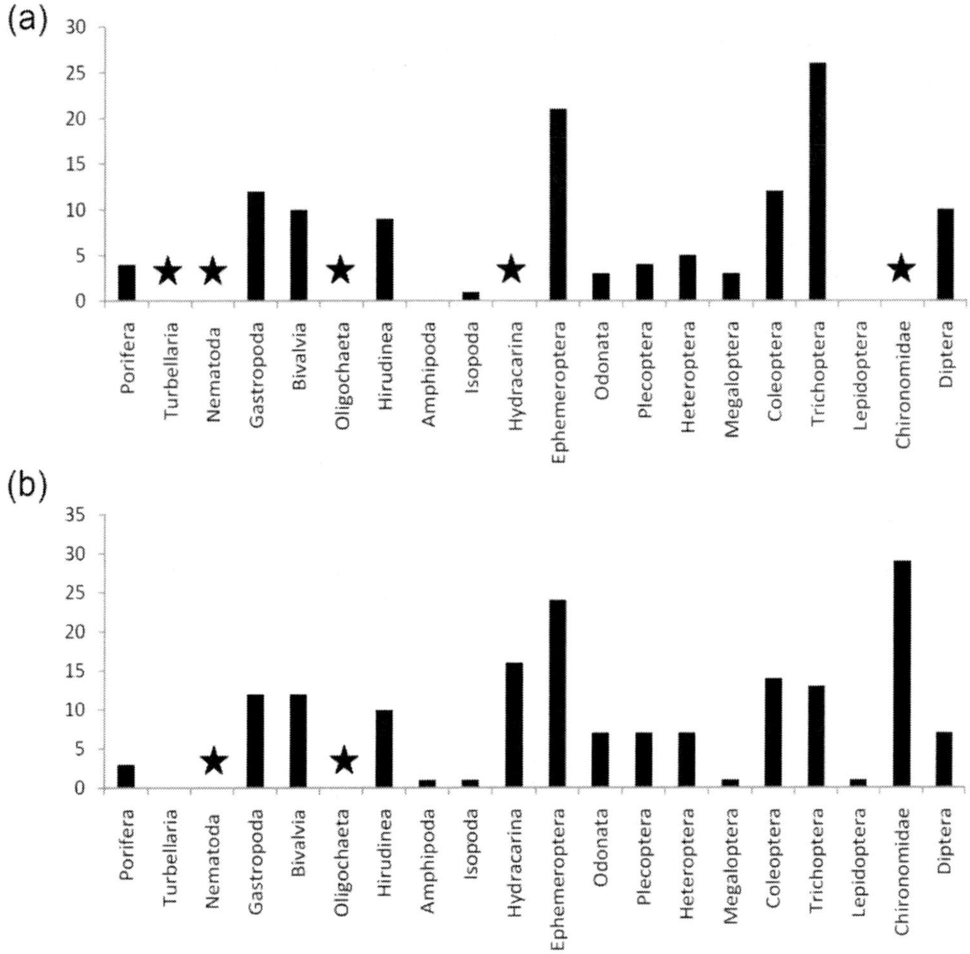

Figure 6. Number of species per taxa group: A – Dataset from the Upper Volga Expedition 2005 (30 sites in the freeflowing section; 90 samples), B – Dataset from the monitoring framework at Tver, Rzhev and Staritsa (3 sites, 25 samples); taxagroups, that were not determined are marked (★).

Considering relative abundances (mean values from all samples from the free-flowing section) Oligochaeta were the dominant group in the freeflowing section of the Volga and constitute 29.7 %, followed by EPT- Taxa, which constituted 16.9 % (Ephemeroptera – 5 %, Plecoptera – 0.9 % and Trichoptera – 11 %), Chironomidae (15.5 %), Mollusca (9.4 %), Hirudinea (5.9 %) and "other taxa" (1.4 %) (Figure 7). A closer view on community characteristics (abundance of main groups in different habitats) is presented in Figure 16. In terms of biomass Mollusca were important organisms in the headwaters of the Volga River. Due to their size they made up about 35 % of the whole biomass (mean value in the freeflowing section = 8.92 g m^{-2}, without *Anodonta* sp. and *Unio* spp.), followed by Hirudinea

(19 %) and Trichoptera (18 %). Although Oligochaeta were the most abundant group, they only comprised 8.5 % and played a minor role considering biomass.

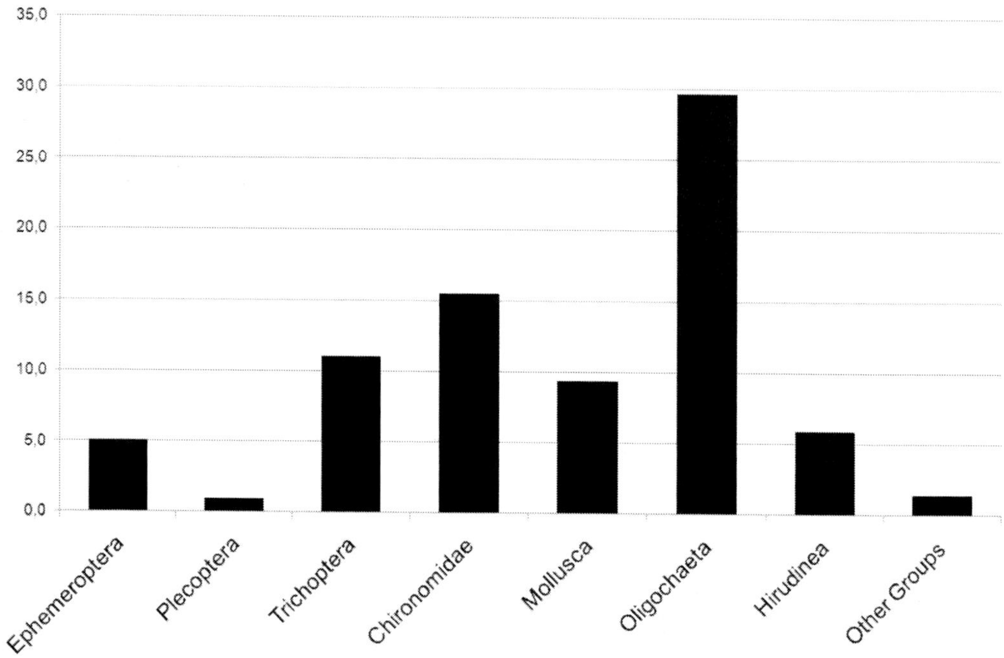

Figure 7. Relative abundances of the main benthic groups in the freeflowing section.

Considering the reservoirs downstream the free-flowing section, macroinvertebrate communities in the Ivankovo, Uglich, Rybinsk, and Gor'kovskoe reservoir were studied in detail (Perova 1998, Kopylov 2001, Bakanov 2003, Shcherbina 2009) and recently a review about the nonindigenous invertebrates in the periphyton and benthos of upper Volga reservoirs was provided (Skal'skaya 2008). In the Upper Volga River about 800 invertebrate species are known today, of which about 280 are Chironomidae (Kopylov 2001). Several Ponto-Caspian species invaded the river within the last 50 years (see Table 3).

Along the middle Volga the benthic communities of the Cheboksary reservoir were rarely observed and only recently studied in detail (Nazarova et al. 2004), while the fauna of the Kuibyshev reservoir is well investigated (Zinchenko 1992, Rozenberg & Vyhristjuk 2008). From the Middle Volga about 600 species are known, with Chironomidae (200 species) and Mollusca (112 species) being the most taxa rich groups (Litvinov et al. 2009).

From the Lower Volga less than 500 macroinvertebrate species are recorded (Litvinov et al. 2009). The highest quantities are found in the same groups as in the upper and middle reaches, such as oligochaet, chironomid and mollusc species, further the ponto-caspian gammarids *Dikerogammarus haemophabes* and *Pontogammarus obesus* are common (Butorin & Mordukhai-Boltovskoi 1978, Litvinov et al. 2009, personal observation). A review about the invasion of invertebrates into the Volgograd reservoir was provided by Filinova et al. (2008).

Table 3. Invasive species that came into the Upper Volga Basin since 1950 (according to Kopylov 2001)

Species	Year	Reference
Polychaeta		
Hypania invalida (Grube)	1989	Shcherbina et al. 1997
Oligochaeta		
Potamothrix vejdovskyi (Hrab)	1969	Semernoi 1974
Potamothrix heusheri (Bretscher)	1969	Semernoi 1974
Psammoryctides moravicus Hrab	1986	Sherbina 1998
Mollusca		
Dreissena bugensis Andrussov	1997	Orlova et al. 2000
Dreissena polymorpha Pall.	1953	Fenyuk 1959
Crustacea		
Heterocope appendiculata G. Sars	1952	Mordukhai-Boltovskoi 1955
Eudiptomus gracilis (G. Sars)	1952	Mordukhai-Boltovskoi 1955
Eudiptomus gracilloides Lill.	1952	Mordukhai-Boltovskoi 1955
Cyclops colensis Lill.	1956	Mordukhai-Boltovskoi & Monakov 1968
Limnosida frontosa G. Sars	1952	Mordukhai-Boltovskoi 1955
Daphnia longiremis G. Sars	1979	Rivier 1986
Daphnia cristata G. Sars	1952	Mordukhai-Boltovskoi 1955
Bosmina longispina Leyding	1952	Mordukhai-Boltovskoi 1955
Bythotrephes longimanus Leyding	1952	Mordukhai-Boltovskoi 1955
Gmelinoides fasciatus (Steb.)	1986	Shcherbina 1996

4.2. Danube

Biodiversity as a result of habitat quality and availability in a given area is hardly measurable in spatial and temporal heterogenous and dynamic ecosystems like large rivers. Its research is therefore mainly restricted by and dependent on the applied methods. Various studies have been carried out at the Danube and their results are highly variable. Russev (1967) for example summarized only 67 species for the Austrian stretch. In contrast Moog et al. (1994) list 897 species of macrozoobenthos for the Danube in Austria including free-flowing sections, backwaters of impoundments and backwaters of the riverside forests based on literature, whereas by far the major share was documented in forested backwaters.

The following part is largely based on results of Joint Danube Survey 2 (JDS 2) at which samples along the entire Danube were taken exclusively from the main channel between August and September 2007 (for details see Graf et al. 2008a). The short study period and the methodological restrictions result naturally in a limited insight of the whole benthic community but this snapshot, however, can be regarded as representative for a characterization of the longitudinal zonation of invertebrate species.

The macrozoobenthic community documented during JDS 2 comprises 411 taxa. The most heterogeneous groups were Diptera (174 taxa) and Oligochaeta (53 taxa) followed by Ephemeroptera (42 taxa), Trichoptera (35 taxa) and Molluscs (Bivalvia 26 taxa, Gastropoda 27 taxa, respectively). Coleoptera (17 taxa), Amphipoda (13 taxa) and Hirudinea (11 taxa) are noteworthy; other groups are less diverse. This overall characteristic in diversity does not change along the three reaches of the Danube, although the number of insects decreases considerably downstream (Figure 8).

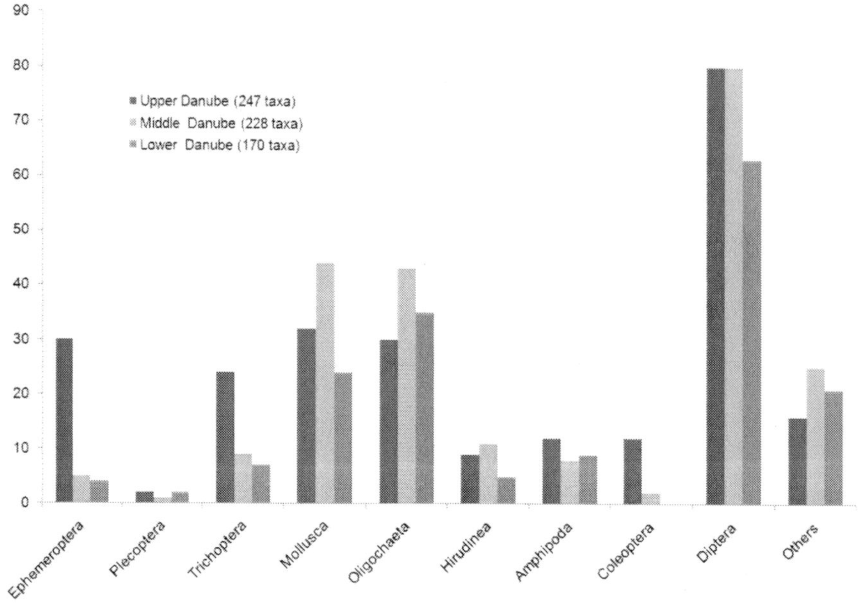

Figure 8. Number of taxa per taxagroup along the different reaches of the Danube.

Rare species found along the JDS 2 are the large burrowing Ephemeroptera *Palingenia longicauda* (Prut River) and the Gastropoda species *Theodoxus transversalis* (Lower Reach). Regarding Amphipoda the invasive species *Corophium robustum* was documented for the first time in Austria (Pöckl, in litt.), *Crangonyx pseudogracilis* was a new record for the Fauna of the Danube (Bernerth & Stein, 2003; Berthold & Kaiser, 2005).

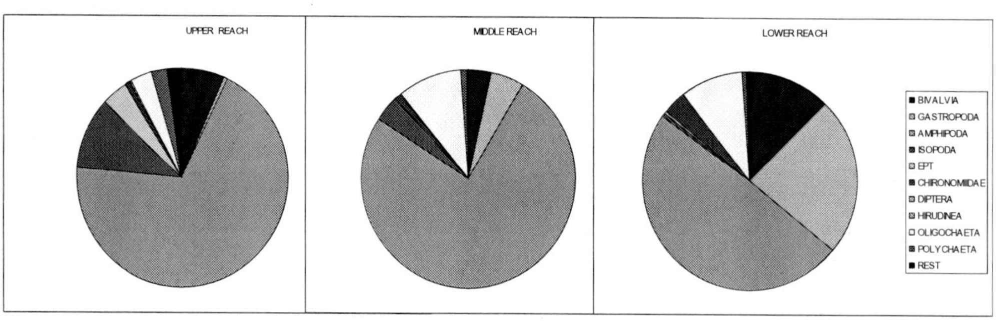

Figure 9. Abundance of Taxa per Taxagroup along the different reaches of the Danube.

Regarding abundance (ind. m^{-2}) Amphipoda are the dominant group in all Danube reaches and constitute up to 75 % and while Isopoda (mainly *Iaera istri*) play an essential part in the Upper Reach and decrease downstream (Figure 9). Oligochaeta and Mollusca can be found in increasing numbers in the Lower Reach. EPT- Taxa (Ephemeroptera, Plecoptera and Trichoptera) were negligible – with the exception of the Upper Reach (above Geising in Germany). Within aquatic Insects only Chironomidae increased downstream and play a major role.

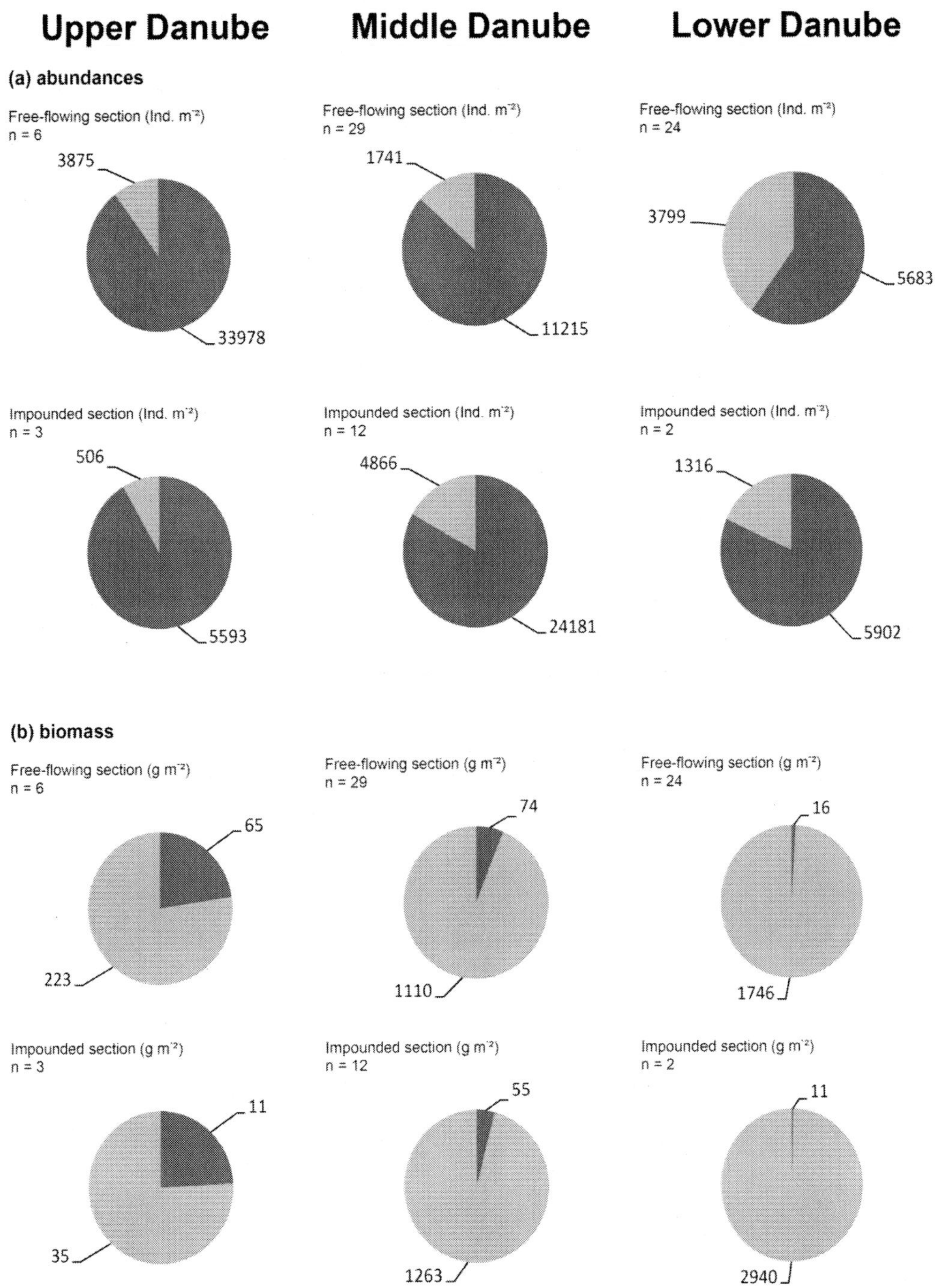

Figure 10. Abundances and biomasses of Mollusca (grey shaded) in comparison to the other taxa (dark shaded). The data is presented for the free-flowing and impounded sections of the upper, middle and lower section of the Danube separately. Number of sampling sites (n) is given for each dataset, at each site 6 Air-Lift samples (0.04155 m^2 each) were taken.

In terms of biomass Mollusca are the most important organisms of the Danube and investigated tributaries (Figure 10). Due to their size Bivalvia make up more than 80% of the whole biomass, followed by Gastropoda (10% to 35%). Looking at the different reaches of the Danube (according to Literáthy et al. 2002) the increasing dominance of Mollusca from the Upper to the Lower Reach becomes evident. Although Crustacea are the most abundant group, they play only a minor role regarding biomass.

The considerable amount of data collected during JDS 2 from the river bottom and from the bank zone illustrated the longitudinal distribution patterns of characteristic Danubian macroinvertebrate taxa that are important regarding biodiversity and protection of species: e.g. the formerly widespread snail *Theodoxus transversalis* is living now in a very restricted section at the Lower Danube.

Neozoa (invasive species) originating from the Ponto-Caspian area, Asia, Australia and North America are a crucial factor influencing the macrozoobenthic community of the Danube. The Danube is a part of the Southern Invasive Corridor (Black Sea-Danube-Main/Danube Channel-Main-Rhine-North Sea waterway), one of the four European most important routes for invasive species (Galil et al. 2007). The river is exposed to intensive colonisation of Aquatic Invasive Species and further spreading in both (north-west and south-east) direct ions throughout the Danube Basin. With few exceptions the neozoa of the Danube belong to Crustacea and Mollusca.

The abundance and percentage of species of neozoa along the three reaches of the Danube are presented in Figure 11, indicating their essential importance for the ecosystem. Due to their tremendous abundance up to 90% within the Upper Reach or even 100% within the Middle Reach of the Danube and approximately 40% of all documented species at the Upper and Middle Reach their impact on each assessment system used becomes evident.

Neozoa dominate the Danube not only locally but they are distributed along the entire stretch. The highest frequency along the Danube River shows the mussel *Corbicula fluminea*. It occurs in 93% of the sites followed by the crustaceans *Corophium curvispinum* (90%) and *Dikerogammarus villosus* (69%). Among the most frequent 15 taxa only *Stylodrilus heringianus* can be regarded as non invasive (Figure 12).

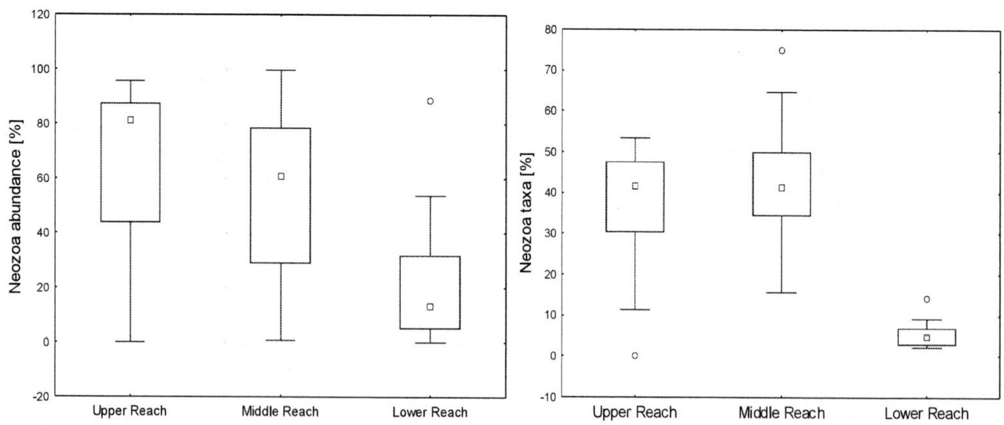

Figure 11. Box & whisker-plots of neozoa abundance and neozoa taxa.

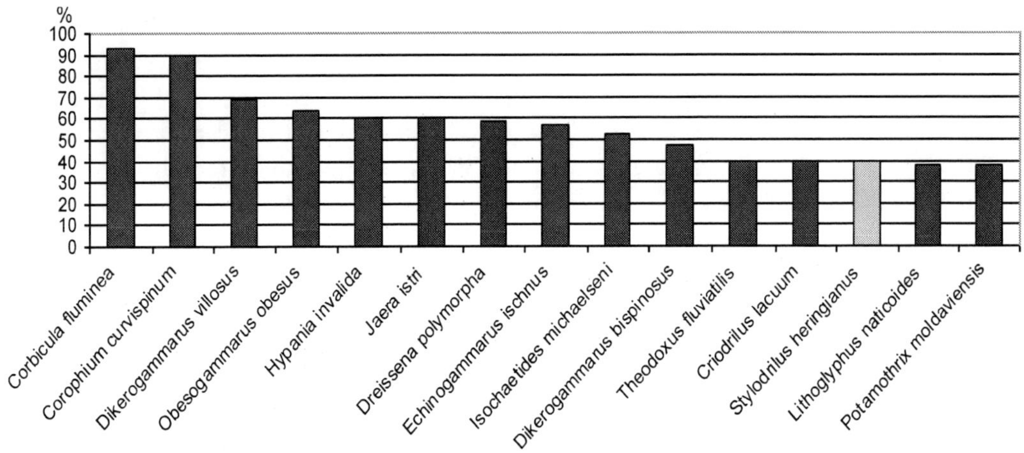

Figure 12. Frequency of taxa along the Danube (free-flowing sections only).

5. Characterisation of the Free-flowing Sections in the Headwaters of Volga River and along the Austrian Danube

5.1. The Free-flowing Section in the Headwaters of Volga River

The "Upper Volga Expedition 2005", which was carried out in August 2005 by scientists from the Russian Academy of Science, Tver State Technical University, Tver State University and the University of Innsbruck, was a detailed investigation of hydrology, hydrochemistry and hydrobiology in the headwater of Europe`s largest river (Kuzovlev & Schletterer 2006).

Catchment analyses resulted that the headwater downstream to the city of Tver amounts 31,300 km^2: about 41.5 % (12,980 km^2) are forested, 2.4 % (760 km^2) covered by mires and 2.1% (668 km^2) of the area are lakes (Surin 1976). Around the Upper Volga Lakes even 66 % are forested and downstream to Eltsy 60 % and the basin downstream to Rzhev still 55 % of the Basin is covered by forests of Pinetum and Betuletum type, belonging to the southern taiga (Klimo & Hager 2001), however the forests of Tver region are increasing constantly within the last decades (Puzachenko, pers. comm.). The mixed forest is dominated by pine (*Pinus sylvestris*), spruce (*Picea abies*), birches (*Betula pendula*, *B. verrucosa*) and on the banks and meadows alder (*Alnus glutinosa*) is dominant. At banks and around islands wood assemblages with bite traces of *Castor fiber* where observed, thus it can be assumed that the European beaver plays a role for the input of wood into the headwaters of Volga River, mainly through tributaries, where a couple of beaver dams are established.

Geologically, the catchment area of the Upper Volga belongs to the Moscow Basin in the Precambrian Russian Sheet, which is composed by crystalline and sediment rocks, as well as metamorphous products (Dolginow & Kropatschjow 1994, Karpunin et al. 1997, Kheraskova et al. 2005, Sahagian et al. 1996) and is composed of east-dipping layers from Devonian to Creatous: the Source and the Upper Volga Lakes lie in Devonian sediments and at Bejshlot the river enters carboniferous layers, which are the main lithological factor in the research area. Some kilometers downstream Tver, approximately at the mouth of River Orsha, Jurassic layers appear, which dominate until Rybinsk reservoir. In this area also some Triassic and very few eroded Permian layers appear, but they are not directly cut by the Volga River (CGI 1971). The surface layers, were formed by glacial activities and the deposited quaternary sediments, consist of conglomerates, sandstones, marls and limestones (Spreitzer 1935) and the main end moraine layers are up to 70 m thick (Gurtz et al. 1999); modern landscape is formed by fluvial activity. Recent soils are mainly loamy-podzolic and in depressions with water accumulation marshy peat-bog-soils occur. The Volga River cuts carbonate deposits, where they lie close to the surface, which is reflected by fossil records made during the expedition in 2005 (e.g. the gastropod *Omphalotrochus* and the brachiopod *Productus*).

Along the 446 km stretch between the source in the Valdaian Hills and Tver there is a difference in altitude of 104 m (= 0.232 ‰). The highest gradient is at Bensky Parog (= 3 ‰) near the village Elzi (Shaporenko et al. 2006). The River Volga between the Upper Volga Lakes and Tver (Figure 13 a-c) has a stream size (order) of six already at the outlet of the Upper Volga Lakes, at Tver the river Tvertsa increases the number to seven (this was assessed with a large scale map 1:100,000). It turned out that there are many small tributaries in the investigated area, but only few large tributaries ones (with stream orders of 5 or 6), namely Zhukopa, Selisharovka, B. Kosha, Vazuza, Derzha, Tma and Tvertsa.

Quantitative sampling of benthic invertebrates was carried out with a bottom grab during an expedition in 2005 (Schletterer 2006) and during subsequent surveys a modified Multi-Habitat-Sampling (MHS) method was applied (Schletterer et al. 2011). In the free-flowing section over 30% of the species assemblage belong of the to the epipotamal zone, and there are also shares of hyporithral and metapotamal elements as well as a substantial percentage of littoral components. Thus the free-flowing section was described as epipotamal zone (Schletterer & Füreder 2010a).

a)

Figure 13. Continued.

b)

c)

Figure 13. (a) Biggest rapid near Eltsy; (b) Volga River at Kokoshkino (photographs: M. Schletterer); (c) The frozen Volga at Tver in March 2006 (photo: A.V. Kurganov).

5.1.1. Biodiversity of the Free-Flowing Section between the Upper Volga Lakes and Tver

During the Upper Volga Expedition approximately every 15 km benthic samples were taken (Figure 14): This was a unique chance to assess biodiversity of the Volga headwaters from the source to Tver` (Schletterer 2006).

In August 2005 a total of 126 species were found in the free-flowing section (river sections D-F), including Trichoptera (26 taxa), Ephemeroptera (21), Mollusca (12), Coleoptera (12), Diptera (10), Bivalvia (10), Hirudinea (9), Heteroptera (5) Plecoptera (4), Porifera (4), Odonata (3), Hydrazoa (1), Megaloptera (3) and Isopoda (1). Chironomidae, Oligochaeta, Turbellaria, Hydrachnellidae, Nematoda, were not determined to species level, thus these groups are handled as one taxon each (Schletterer & Füreder 2010a).

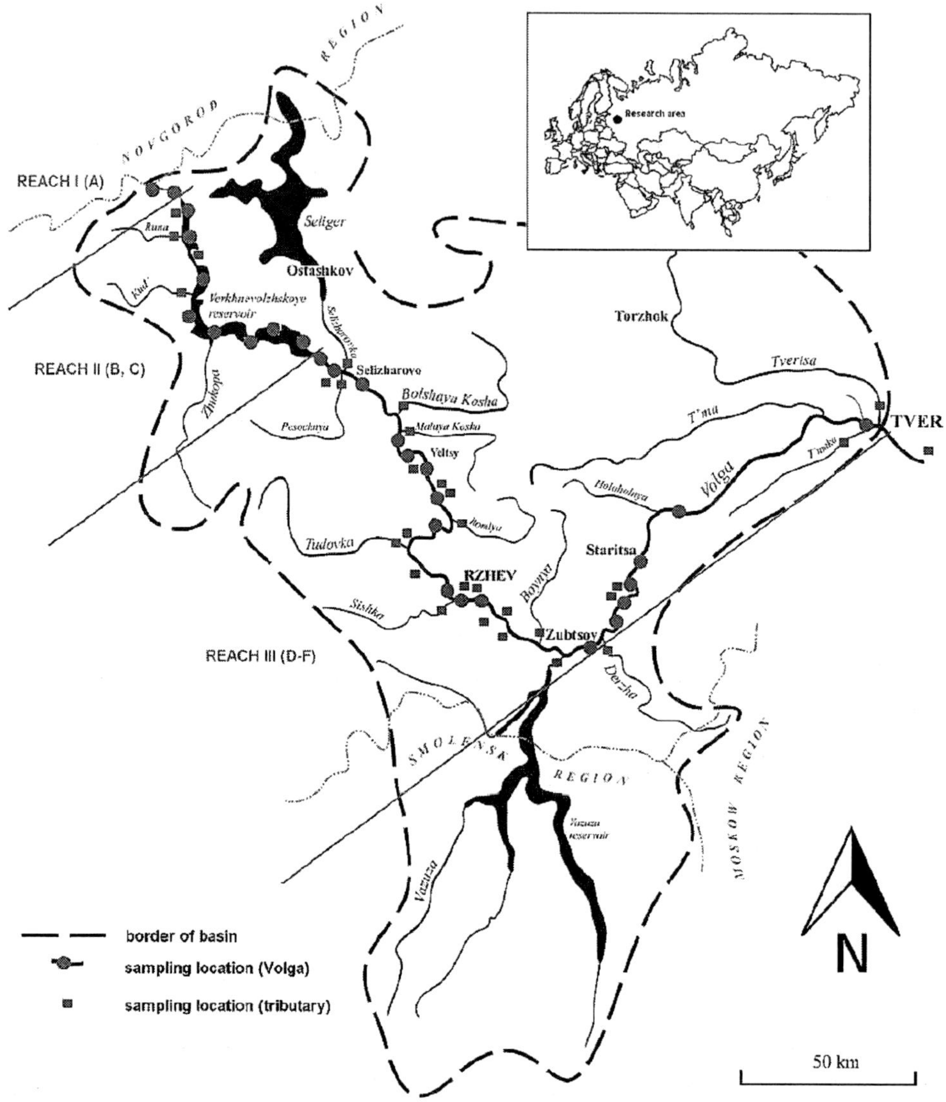

Figure 14. Research area in the headwaters of the Volga River (from: Schletterer & Füreder 2010a).

The mean abundance of macroinvertebrates was 3,496 ind. m^{-2}, highest densities were observed near Petunovo (14,760 ind. m-^{2} in a single sample; three replicates pooled = 9,587 ind. m-^{2}). In Figure 15 the frequency of the 15 most common taxa is presented. Overall Chironomidae (Diptera) were the most abundant insects, with means ranging from 20.3 to 36.7 % of the total community in the free-flowing section. A comparison of settlement densities in different substrate types is provided in Fig. 16. With 42 to 84 species, the free-flowing section had highest species numbers. Fresh weight was increasing along the continuum from 11.76 to 26.76 g m^{-2}. The natural free-flowing section and its tributaries comprise a diverse epipotamal fauna. Downstream from the Upper Volga Lakes, the functional feeding groups followed the predictions of the River Continuum Concept: in the upper reaches the community in the free-flowing section is composed of filter feeders, followed by grazers, predators, detritivores and shredders while further downstream detritivores became slightly dominant (Schletterer & Füreder 2010a).

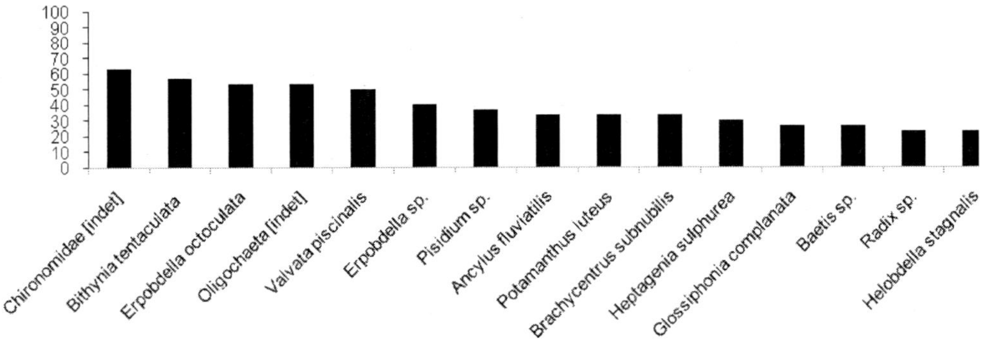

Figure 15. Frequency of the 15 most common taxa at 30 sites in the free-flowing section.

The Shannon index was about 2.3 and the Evenness characterized the communities as well structured (Table 4). High similarity values (Sørensen similarity index) were observed in the free-flowing section between the Upper Volga Lakes and Tver` (range 0.44 - 0.56). Analysis of beta-diversity showed that 50 % of species were common to the main river and its tributaries. This observation underlines the importance of the tributaries for the whole river basin. The river network is highly interconnected and several typical epipotamal elements also find suitable habitats in the tributaries.

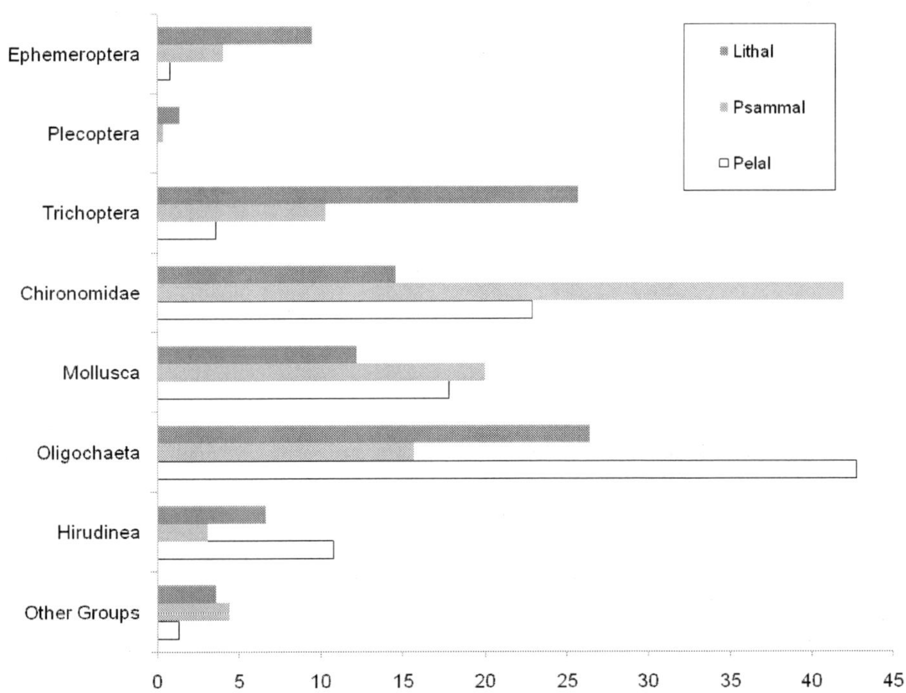

Figure 16. Shares of benthic main groups within different habitats (lithal, psammal, pelal) in the free-flowing section.

Table 4. Community metrics and diversity indices at the three sections of the free-flowing section (from: Schletterer & Füreder 2010a)

	Section D	Section E	Section F
River section	Bejshlot – Eltsy	Eltsy – Rzhev	Rzhev – Tver'
River km	105–153	153–261	261–446
Nr of sites (samples)	12 (36)	7 (21)	10 (30)
Mean nr. of taxa per sample	15	18	24
Total richness	77	51	102
Abundance (Ind. m^{-2})	1565 ± 678	3668 ± 3416	3431 ± 1608
Biomass (g m^{-2})	24.35 ± 11.39	22.12± 15.67	26.76 ± 9.77
Shannon index	1.53 ± 0.69	1.48 ± 0.85	1.80 ± 0.57
Evenness	0.70 ± 0.20	0.69 ± 0.30	0.64 ± 0.17

5.1.2. The Diverse Fauna of the Potamon in the Headwaters of the Volga River, in Comparison with Historical Data of the Danube River

Within the expedition in 2005 and a subsequent monitoring programme, 29 mayfly species were recorded in the headwater of Volga River (Schletterer & Füreder 2010b). Some of these species are typical mayflies in the large European rivers: The species *Potamanthus luteus* is known to be distributed throughout Europe (Illies 1978) and is a common species in East-European lowland rivers (Obrdlik et al. 1979, Soldan et al. 1998). Also *Heptagenia sulphurea* is a common grazer in lowland rivers, like in the Volga and Danube, that prefers stony substrates (Banning et al. 1990). The burrowing mayfly *Ephoron virgo* formed impressive mass-occurences in former times (Schoenemund 1930), however it became rare in central European rivers and only recently re-appeared at the lower Rhine as well as in the Danube (Kureck 1992, Bathe 1997, Literáthy et al. 2002). In the headwaters of the Volga River this species forms such mass emergences regularly in the middle of August, with a main flight activity between 21:45 and 22:30 (Schletterer & Füreder 2010b). Europe`s largest mayfly *Palingenia longicauda* was found in 1924 by Behning in the Middle Volga and it was listed by Russev (1967) for the Danube, but it became rare within the last century. According to local people it could be also present in the headwaters of Volga River near Petilovo upstream Tver (Schletterer & Füreder 2010b). Concerning the Danube, during the JDS 2 larvae were found at the Prut River, a tributary to the Danube in Ukraine (Graf et al. 2008a). However, nowadays only one single population seems to have survived at a large scale in the Tisza River in Hungary (Tittizer et al. 2008). The very rare mayfly *Behningia ulmeri* Lestage 1929 was discovered in the beginning of the last century in the Volga upstream Rybinsk (Behning 1924, Ulmer 1924), the family Behningiidae was set up by Tshernova (1938). The larvae are living in large rivers on sandy substrates and are adapted for burrowing in the sand (Ulmer 1924). Further records were made in the Ural (Behning 1932), Dniester (Motas & Bacesco 1937), Amur (Tshernova 1938), Vistula and Warta (Keffermüller 1957, 1959, 1963), but it seems that the species disappeared nowadays due to river regulation and pollution (Kłonowska-Olejnik 2009). Another very rare mayfly – *Prosopistoma pennigerum* – was recorded during monitoring activities in the headwaters of River Volga for the first time in Russia (Schletterer & Kuzovlev 2007). This species was found in the 19th century regularly in European watercourses but disappeared in the 20th century almost completely (Schletterer &

Füreder 2009, and references therein). Thus the records in the headwaters of the Volga River were a unique chance to assess habitat specifics of the species. Entomological notes on all described species of the mayfly family Prosopistomatidae were recently summarized by Schletterer & Füreder (2008). Three key factors, i.e. ecological preferences were defined for the European species *P. pennigerum*: substratum (lithophilous), current (rheophilous) and temperature regime (potamophilous) (Schletterer & Füreder 2009).

The nymphs of Plecoptera are usually living in rhithral water bodies, i.e. cool or cold running waters, thus in the headwaters of the Volga River, that emerges already in the lowland ("potamo-crenal" sensu Schletterer & Füreder 2010a), only a few stonefly species were found. In the headwaters of Volga River *Amphinemura sulcicollis* is the dominant stonefly in spring (emergence likely starts in the middle of May). In summer (August) larvae of the leuctrid *L. fusca* were found in the Volga River as well as in tributaries in high densities. *Nemoura cinerea* as well as *Nemurella pictetii* have a wide ecological range, anyway both are typical for the slightly acidic waters emerging from fens and mires. These species were found to occur together in tributaries (e.g. at some sites along Tudovka) and the source of Volga River, but *N. cinerea* was always the predominant species. The circumpolar species *Diura bicaudata* was found at the source of River Tudovka as well as in a potamal section of the Volga River. Also from Fennoscandia this species is known from streams, rivers and lakes (Lillehammer 1988), in contrast the species is known in Central Europe (Austria) only from few crenal or hypocrenal sites (Graf & Schmidt-Kloiber 2008, Graf et al. 2009). Two *Isoperla* species were found: In the Volga River *Isoperla* cf. *difformis*, which prefers rivers with a fine-grade substratum beneath the stones in the stream bed (Lillehammer 1988), was found. The second species is *I.* cf. *obscura*, which is a sensitive stonefly of large rivers and became rare nowadays throughout Europe (Upper Danube, Drau, Inn). Also *Xanthoperla apicalis* is known mainly from larger rivers (Lillehammer 1988) and became rare in Europe (Graf & Schmidt-Kloiber 2008).

Diversity in Plecoptera regarding the river Danube seems to decline since about 60 years. Many potamon species have not been found over about 50 years like *Isogenus nubecula* (last date of record 1957), *Brachyptera braueri* (1906), *Taeniopteryx araneoides* (1902), *Brachyptera trifasciata* (1925), or even since the 19th century like *Oemopteryx loewii* (1889, Regensburg). The combination of channelisation in 18th century and pollution and damming during the 20th century seemed to be a too deadly cocktail. Surprisingly, *Xanthoperla apicalis* recovered and can nowadays be found in the middle of Vienna (Graf 2010) and around Bratislava again. *Isoperla obscura* survived in the upper Danube at Engelhartszell, populations at Vienna were found until 1958 (Graf 1999).

The caddisflies were found to be very diverse in the free-flowing section in the headwaters of the Volga River. Within the Polycentropodidae several species were found, e.g. *Neureclipsis bimaculata* which construct typical nets to catch its prey. *Cyrnus trimaculatus*, a species of the river margin as well as backwaters, is described to be highly sensitive to environmental contaminants, thus this indigenous species of large European rivers is an important measure for ecological recovery (Geest et al. 2000). *Psychomyia pusilla* (Psychomyiidae) is a typical grazer in the hyporhithral and epipotamal and was common in sites dominated by lithal between Rzhev and Tver. The Hydropsychidae are a species rich group in running waters with a specific zonation, i.e. *H. bulgaromanorum* is restricted to large rivers (Lepneva 1970, Botosaneau & Malicky 1978, Pitsch 1983, Graf et al. 2008b).

But also the other benthic groups were diverse and faunistic highlights, i.e. characteristic potamal species, were found as well: Considering Porifera, *Trochospongilla horrida* was found to be the most common species in the headwaters of the Volga River (Schletterer & Eggers 2006). This species is also known from other large rivers like the Rhine (Gugel 2000), Elbe (Eggers 2006) or Danube (Literáthy et al. 2002). Three of the Odonata in the headwaters of the Volga River are "strictly protected species" of European interest (*Aeshna viridis, Ophiogomphus cecilia* and *Stylurus flavipes*). The saucer bug (*Aphelocheirus aestivalis*) was common in most large European rivers as a typical component of the bentic community in clean waters with highly structured mesohabitats (Statzner 1979), but became rare due to anthropogenic impacts, especially pollution (Damgraad 2005). In the headwaters of the Volga River the species was abundant in stony locations that were rich in sumerged macrophytes. Also the findings of *Unio crassus, Margaratifera margaritifera* (FFH Annex II & IV species, EC 1992) – the latter one is restricted to tributaries (Schletterer et al. 2011) – and *Unio tumidus* as well as the other molluscs are remarkable, because in European Rivers the mollusc fauna decreased significantly within the last century (Jueg & Zettler 2004). In the Danube, among molluscs *Unio tumidus* and *Unio pictorum* are still widespread species. On the other hand *Theodoxus danubials* and *T. transversalis* (due to the spread of *T. fluviatilis*) as well as *Unio crassus* spp. are restricted to small stretches along the Danube and they might even become extincted before we even know their exact identity (Graf et al. 2008a). Especially in well-vegetated freshwater habitats aquatic beetles are diverse: worth mentioning are *Dytiscus latissimus*, that was found in a meandering section of river Tudovka, *Hydrous piceus* which occurred in dense macrophyte stands in the Volga and also the genus *Normandia*. The Annex II and IV species *Dytiscus latissimus* (EC 1992) is known mainly from dystrophic stagnant water bodies, however also a few records are known from slow flowing rivers (Rau 1888), e.g. Galewski & Tranda (1978) mention old records from the Vistula.

5.1.3. Definition of Reference Conditions for Lowland Rivers

Benthic invertebrates are important parameters to assess river health, thus different indices for the assessment of the ecological status were used and evaluated. The European Water Framework Directive (WFD), provided an important regulation for the assessment of the ecological quality of surface water bodies, based on biological components. The ecological status is described as "deviation from the reference condition", however for large lowland rivers there are still difficulties to identify reference conditions, because reference sites in Central Europe are lacking.

Dynesius & Nilsson (1994) described the Volga River, as well as most European rivers, as strongly affected water body. The fragmentation of Volga was classified with "4", i.e. only 0 to 24% of the main-channel are without dams and there are also dams in the catchment of the largest tributary. Further, flow regulation was described by reservoir live storage (34), reservoir gross capacity (75) and irrigation consumption (2) for the entire river system, expressed as the percent of its VMAD (virgin mean annual discharge = 8050). Despite large river systems might have unaffected sections or tributaries, Dynesius & Nilsson (1994) concluded, that intact sections or tributaries can not be handled as complete rivers, because they have no connection to the sea (i.e. an estuarine flora and fauna is missing). Contrary, our opinion is that a pristine headwater or intact tributaries, are important for aquatic ecology to understand natural processes and dynamics in a large river system (e.g. Schletterer & Füreder 2009). Due to minor changes in river morphology, catchment conditions (high amount of natural forest) and a low population density the headwaters of Volga River enable the definition of reference conditions for lowland rivers, according to the EU-Water Framework Directive (Schletterer & Füreder 2010a).

Thus, with the data from the intact free-flowing section in the headwaters of Volga River, the relations of bioassessment indices with natural environmental factors were investigated: Saprobic Index (Zelinka & Marvan 1961), SPEAR indices (Liess et al. 2008, Beketov & Liess 2008, Beketov et al. 2009), and other macroinvertebrate community indices and basic environmental factors (see Schletterer et al. 2010). This allowed the definition of reference conditions, i.e. the definition of reference taxa (Schletterer & Füreder 2009) and the ecological status classification for SI and $SPEAR_{pesticides}$ indices (Table 5). Saprobic reference conditions were found to be beta-mesosaprobic. Furthermore, the results indicated that the index $SPEAR_{pesticides}$ is potentially applicable across different types of watercourses as well as across different biogeographical regions in Europe showing similar boundaries of ecological status classes (Schletterer et al. 2010). The proposed boundaries for the saprobic index are comparable to the Austrian classification for lowland watercourses (High $\leq 2.00 <$ Good $\leq 2.40 <$ Moderate $\leq 2.80 <$ Poor $\leq 3.20 <$ Bad, Ofenböck et al., 2007). Besides, it is also comparable to the German system, where type-specific boundaries between the classes High and Good were defined as ≤ 2.00 (for stream type 12, rivers with a high degree of organic matter), ≤ 1.85 (for type 15, mid-sized sand bottom streams) and ≤ 2.05 (for type 21, lake outlets) (Rolauffs et al., 2003).

Table 5. Class boundaries for the Saprobic index and $SPEAR_{pesticides}$ (according to Schletterer et al. 2010)

Status class	Saprobic index	$SPEAR_{pesticides}$
High	≤ 2.18	≥ 0.44
Good	2.18 – 2.51	> 0.30
Moderate	2.52 – 2.81	> 0.22
Poor	2.82 – 3.27	> 0.11
Bad	> 3.28	≤ 0.11

5.2. A Cross-Section of the Free-flowing Section of River below Vienna

The River Danube east of Vienna (Figure 17 a-c) has a stream size (order) of nine, but the free-flowing section (Figure 17c) resembles middle reaches (stream order six) along the river continuum (Vannote et al. 1980). The value of six agrees with zonation concept of Illies & Botosaneanu (1963), whereas the high stream order (nine) is simply a consequence of the inflow of another large river, the Inn.

In the free-flowing section 30% of the species assemblage belong of the to the epipotamal zone, 20% belong to the hyporhithral zone (Illies & Botosaneanu 1963) and the remainder belong to seven other zones. Therefore this particular section of river represents a transition zone between the hyporhithal and epipotamal zones (Moog et al. 2000).

Quantitative sampling and methods of analysis were applied to a transect across the free-flowing main channel below Vienna at river-km 1889.9 (Figure 19a; Figure 17c). Here the Danube is about 380 m wide, with a depth ranging from 3.1 to 4.6 m at a mean discharge of 2170 m^3 per second (Fesl et al. 1999). The riverbed is composed of rocks, stones and gravel, with occasional patches of sand and silt. The riverbed was sampled to a depth of 0.26 to 11.1 cm, using a modified Petersen grab (Humpesch & Elliott 1990).

a)

b)

c)

Figure 17. (a) Freudenau power-station downstream of Vienna (river-kilometre 1921), view upstream; (b) River Danube ice covered in Vienna on 6 February 2006, showing the Danube Tower, the Vienna Conference Centre and part of the UN-City; (c) Free-flowing stretch of the River Danube near Wildungsmauer (river-kilometre 1895), looking upstream (all photographs: U. Humpesch).

5.2.1. Biodiversity of the Macrozoobenthos Communities in Relation to Environmental Variables at the River Cross-Section below Vienna

Important Environmental Variables at the River-cross Section

Six major environmental variables were extracted from a principal component analysis, of which five are important in determining the species composition and relative abundance of invertebrates across the river channel (Table 6). The sixth was water temperature ranging between 2 °C and 19.5 °C for the study period.

Table 6. Six hydrophysical variables at four sampling sites in a cross-section of the River Danube at river-km 1889.9 (from Fesl 2002).

Variable		Site 1	Site 2	Site 3	Site 4	Sites	Dates
Water velocity (cm sec^{-1})	*v*	182.0	206.2	197.4	139.6	*	*
Grain size (mm)	*Q2*	59.5	28.9	40.7	27.9	*	*
Substratum heterogeneity	*So1*	1.48	1.58	1.67	1.75	*	
Sediment turnover	*UI*	1.5	3.5	2.6	2.1	*	*
Pore space (%)	*PS*	30.8	33.3	31.5	25.6	*	*
Water temperature (°C)	*temp*		10				

Samples were obtained monthly between September 1995 and August 1996. Location of the sites are shown in Figure 3. *, significant differences ($P < 0.05$) between sites and dates.

At site 4 of the cross-section, close to the left bank, mean water velocity is significantly lower than at the other sites (Table 6). Temporal differences in the water velocity, highest in May and July, at a period of high water result from snowmelt in the Alps. Substratum composition also differed along the cross-section of the river. The right bank at site 1, is artificially stabilised with large boulders, the median of the natural grain sizes is significantly higher compared with other sites of the transect. The smallest particles occur near the left bank behind a sill of sediment; in the mid-channel, at site 3, particles are intermediate in size. Mean grain size also varies significantly through the year. The heterogeneity of the sediment composition increases slightly from the right to the left side of the river channel, and temporal variation is low.

Sediment turnover differs between sites and temporal variations resembled those of discharge, with a lower overall sediment movement during winter and higher turnover in late spring and summer. Sediment deposited close to the left bank is characterised by a lower pore space compared with the other sampling sites, and this might be attributed to the lower water velocity and sedimentation of small particles. Through the year, the lowest pore space was found in July.

Taxonomic Composition and Abundance of Species

A total of 164 species were found at the four sites along the transect (Table 7), including 81 Chironomidae (mainly Orthocladiinae) and 33 Oligochaeta (mainly Naididae). Site 4 near the left riverbank has 142 species including 68 species of Chironomidae (29 of these occur only at site 4) and 30 species of Oligochaeta (2 occur only at site 4). Another seven major groups of consist of 50 species. Site 4 has the highest number of Trichoptera (10 out of 11 species at all sites), Ephemeroptera (9/11), Bivalvia (8/8), Gastropoda (7/8), Crustacea (6/7) and Hirudinea (2/2); two species of Bryozoa occur at three of the sites (Table 7). Altogether, site 4 has by far the greatest number of species (47) that were not found at any of the other sites. The species-accumulation curve for the the free-flowing main channel below Vienna at river-km 1889.9 for the whole period was best described by a power function (Figure 18). The same function was also applicable to the species-accumulation curve obtained by Jackknife estimates. Due to the high number of rare species, Jackknife techniques estimated 211±14 species, compared to 164 actually found (Table 7).

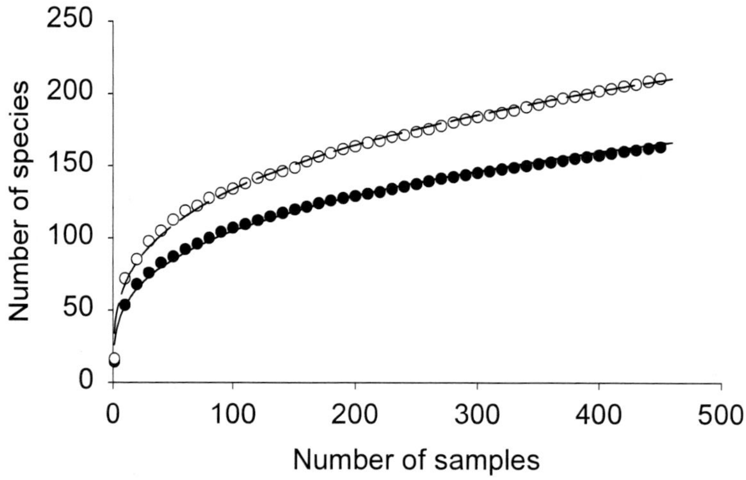

Figure 18. Species-accumulation curve for observed numbers (full symbols) and Jackknife estimates (open symbols) for the whole study area over the sampling period. Fitted power models are indicated by the lines (observed: $Y = 25.3\ X^{0.31}$; $r^2 = 0.98$; $P<0.001$; Jackknife estimates: $Y = 34.0\ X^{0.30}$; $r^2 = 0.98$; $P < 0.001$).

Table 7. Total numbers of species (S), numbers found at each site in the river the cross-section below Vienna; bold numbers in parentheses indicate species occurring at one site only, and no other site. Data for Chironomidae taken from Fesl (2002), for Oligochaeta from Fesl & Humpesch (2003) and for other groups from Fesl & Humpesch (2006a). Location of the sites are shown in Figure 19a)

Group	S	Site 1	Site 2	Site 3	Site 4
Chironomidae	81	40 (**3**)	37 (**8**)	27 (**2**)	68 (**29**)
Oligochaeta	33	28 (**1**)	24 (**1**)	25 (**1**)	30 (**2**)
Ephemeroptera	11	2 (**1**)	5 (**1**)	3	9 (**5**)
Trichoptera	11	7 (**1**)	4	4	10 (**3**)
Bivalvia	8	5	2	3	8 (**2**)
Gastropoda	8	4 (**1**)	1	2	7 (**4**)
Crustacea	7	4	6 (**1**)	4	6 (**1**)
Bryzoa	3	2	2	1 (**1**)	2
Hirudinea	2	1	1	1	2 (**1**)
Total number of species	164	93 (**7**)	82 (**11**)	70 (**4**)	142 (**47**)

Site 4, with 142 species, had the lowest water velocity and grain size. In comparison, site 1 near the right riverbank has the largest median grain size and lowest sediment turnover (Table 6) and has fewer species (93) than compared with site 4, but slightly more than at mid-channel sites 2 and 3, except for Ephemeroptera (Table 7). Seven out of 93 species occur only at site 1, 11 out of 82 species occur only at site 2, and 4 out of 71 occur only at site 3; of these 22 species with restricted distributions, 16 are chironomids and oligochaetes, and two are ephemeropterans.

In terms of total numbers of individuals per m^2 of riverbed, oligochaetes are the most numerous [*Chaetogaster diastrophus/setosus* (Svetlov) and *Nais elinguis* Müller were most abundant], followed by chironomids [*Cladotanytarsus cf. vanderwulpi* (Edwards) and *C.* cf. *daviesi* Bilyj & Davies were most abundant]. These two groups accounted for 95% or more of the total numbers of all individuals of all species found at each site (Figure 19b). The overall mean abundance of all species was highest at site 4 (11,146 individuals per m^2), being double the mean number at site 1 and roughly 10 times the number at site 3; site 2 had on average only 623 individuals per m^2 (Table 8).

Table 8. Seven community characters at four sites in the river cross-section, estimated by Jackknife procedures: richness, diversity, eveness, width (near 0 = low, 1= high), overlap (0 = no, 1 = complete), aggregation (-1 = regular, 0 = random, +1 = clumped distribution), ß-diversity along a gradient (near 0 = low, 1 = maximum change); also given are the values for persistence (averaged correlation coefficients, 0 = complete change, 1 = no change) and the geometric mean abundances

Communitiy characters		**Site 1**	**Site 2**	**Site 3**	**Site 4**
Species composition	Richness (*S*)	113	107	91	182
	Diversity (*H'*)	1.41	0.98	1.18	2.29
	Evenness (ε)	0.13	0.09	0.12	0.17
Spatial resource utilisation	Width (β)	0.06	0.03	0.02	0.02
	Overlap (α)	0.17	0.11	0.12	0.13
	Aggregation (I_p)	0.67	0.72	0.70	0.74
Temporal and spatial communitiy turnover	Persistence (τ)	0.66	0.55	0.53	0.60
	β-diversity (*BD*)	0.05	0.09	0.09	0.06
Mean abundances (per m^2)		5314	623	1195	11146

For the other 50 macrozoobenthic invertebrates that have been identified, Crustacea were relatively most abundant, particularly *Jaera istri* Veuille (35%) and *Corophium curvispinum* Sars (32%), followed by a bryozoan, *Fredericella sultana* (Blumenbach) (4%), and Trichoptera, particularly *Hydropsyche contubernalis* McLachlan and *Psychomya pusilla* (Fabricius) (3%). Mean relative abundances of 1% or less also occur in Bivalvia, Gastropoda, Ephemeroptera and Hirudinea, indicating the presence of many rare species. Total densities for all species show significant differences between sampling sites and great fluctuations through time, with peaks in November and March and lowest densities in July after floods caused by snow-melt in the Alps.

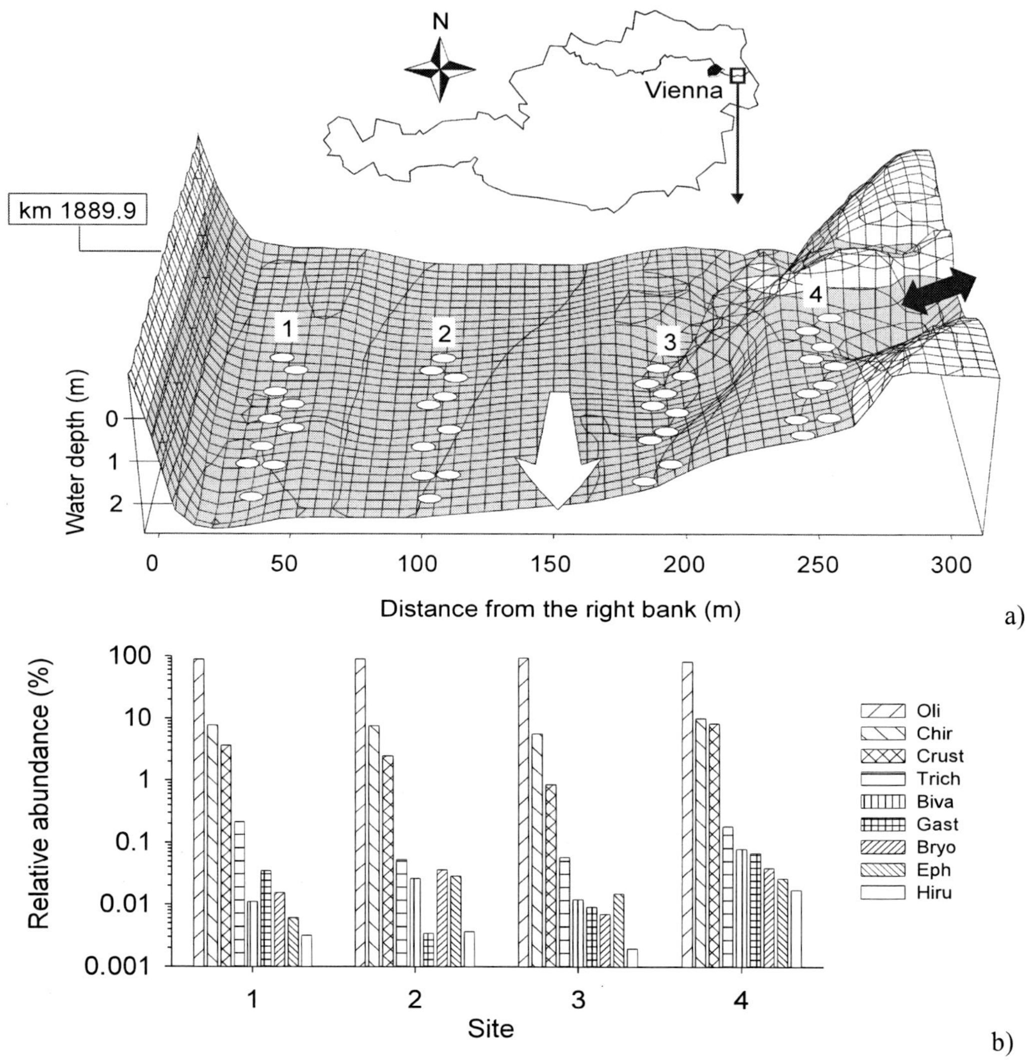

Figure 19. (a) Grid-map of the study area at low water level (discharge 1182 m^3 per sec) at river-km 1889.9 of the Danube, in Austria, east of Vienna. Circles indicate the possible position of the ten random samples within the four sites along the cross-section (sites 1, 2, 3 and 4 respectively at 30 m, 100 m, 200 m and 250 m from the right bank). Samples were obtained monthly between September 1995 and August 1996. White arrow, direction of flow; black arrow, connected to a backwater area (from: Fesl 2002); (b) Relative abundances (log scale) of nine taxonomic groups that were determined to species level. Oli, Oligochaeta; Chir, Chironomidae; Crust, Crustacea; Trich, Trichoptera; Biva, Bivalvia; Gast, Gastropoda; Bryo, Bryozoa; Eph, Ephemeroptera; Hiru, Hirudinea.

Community Characteristics

When comparing the species composition characteristics, Site 4 in the cross-section close to the left riverbank species richness (actual number of species present) and *H'* diversity show higher values than the other sites. *H'* diversity was explained mainly by the dominance structure (expressed as evenness, a measure of the number of species related to numbers of individuals) rather than by species richness (Table 8). Actual observed values of species richness at site 4, with 142 species, is double the observed richness of site 3, which had fewer species (91) than the observed numbers at site 1 and 2. Spatial resource width indicates the portion each species takes from the available space, spatial resource overlap indicates the mutual utilisation of the space of different species, and the aggregation coefficient describes the spatial distribution of single species. Mean spatial resource width is significantly higher at site 1 in the cross-section whereas both mean spatial resource overlap and mean aggregation of the species are not significantly different between sites. Mean spatial overlap is weakly but significantly correlated to mean spatial resource width (Table 8). The values of the characteristics for temporal and spatial community turnover show that there are no differences between the sampling sites, but temporally the lowest persistence values in July and August are significantly compared with those found from October to March. The mean β-diversity is generally low and at site 1 significantly less than at sites 2 and 3. It declines with increasing species richness according to a logarithmic function as a consequence of many rare species.

Quantitative Relationships between Nine Community Attributes and Six Environmental Factors

Redundancy Analysis (RDA) was applied to data obtained at each sampling site for each sampling date (Figure 20). Four RDA-axes explained 99% of the relation between nine community attributes and six environmental factors, 80% of which were explained by the first two axes. Among the relationships between hydrophysical variables and the four axis, sediment turnover shows a strong correlation to the first axis. Sediment heterogeneity is also correlated to this axis. Both grain size and pore space are negatively related to the second axis and sediment heterogeneity is positively related to the second axis.

Relationships between environmental variables and macrozoobenthic communities at the four sampling sites indicate that increasing sediment turnover (at mid-channel sites) coincides with decreasing species richness and community persistence, but coincides an increase in mean spatial aggregation of the species (Figure 20). The latter is also positively related to sediment heterogeneity. A decrease in species richness also coincides with an increase in water flow. Abundance was negatively correlated to sediment turnover and pore space (indicated by bold numbers in Table 9).

Table 9. Correlation coefficients between six hydrophysical variables and nine community characteristics after the RDA, where *v* is water current, *temp* is water temperature, *Q2* is the median of grain-size distribution, *So1* is the coefficient of sorting (heterogeneity of the sediment), *PS* is pore space and *UI* is sediment turnover (habitat instability)

	V	*temp*	*Q2*	*So1*	*PS*	*UI*
Abundance	-0.37	-0.26	-0.22	0.00	**-0.47**	-0.42
Richness	**-0.51**	-0.07	-0.14	-0.08	-0.30	**-0.48**
Diversity	-0.14	0.27	0.15	-0.29	0.07	-0.18
Evenness	-0.02	0.26	0.23	-0.35	0.17	-0.08
Resource width	-0.17	-0.18	0.27	-0.41	0.10	-0.41
Spatial overlap	0.03	-0.13	0.24	-0.32	0.02	-0.28
Aggregation	0.17	0.23	-0.23	**0.52**	0.01	**0.50**
Persistence	0.06	-0.47	0.01	-0.22	-0.12	**-0.47**
β-diversity	0.30	0.35	-0.04	0.34	0.08	0.44

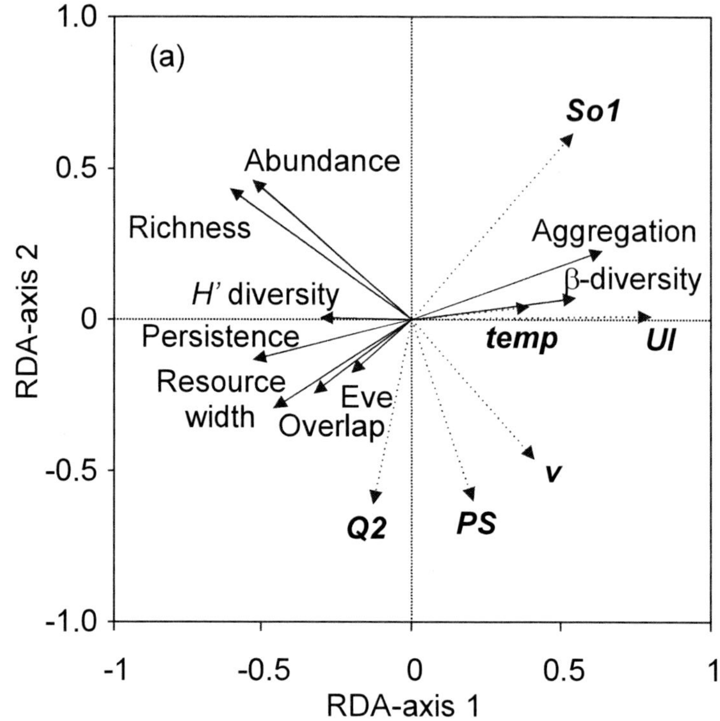

Figure 20. Biplots from Redundancy Analysis (RDA) showing the first and second RDA-axes correlated with nine named community attributes (——;Eve = evenness) and six environmental variables (- - -; *v*, water velocity; *Q2*; median grain-size; *Sol*, coefficient of sorting; *temp*, water temperature; *UI*, index of sediment turnover; *PS* = pore space).

The analysis by RDA suggest, that variation in biodiversity was determinded mainly by eveness rather than by species richness. This suggests that the cross-section of river is saturated with species, leaving no room for new entrants. A decrease in biodiversity and number of individuals coincided with increased water velocity at some sites and, with increased movement of sediment, the spatial distribution of species decreased in combination with a decline in overlap of areas colonised by single species.

5.2.2. Community Organisation of Larval Chironomid (Diptera) Assemblages and Environmental Factors in a Large River: Prediction and Validation of their Interrelationships

Comparable quantitative, longer term studies of the natural variation of the macrozoobenthos in running waters are, in general, scarce and for larger rivers, like the Danube, in particular, because most investigations over a longer period have focussed on specific man-made impacts such as river regulation or the construction of impoundments.

The results from the first year of the investigation (1995 to 1996) stressed the importance of the environmental factors on the selected community characteristics of the macrozoobenthos assemblages: chironomids (Fesl 2002), oligochaetes (Fesl & Humpesch 2003) and the macrozoobenthos community as a whole (Fesl & Humpesch 2006a). Therefore the environmental variables of the following years, 1996 to 1998, were used to predict the community characteristics of the richest taxonomic group, the chironomid larvae, and to compare the estimates with the actual values (Fesl & Humpesch 2006b).

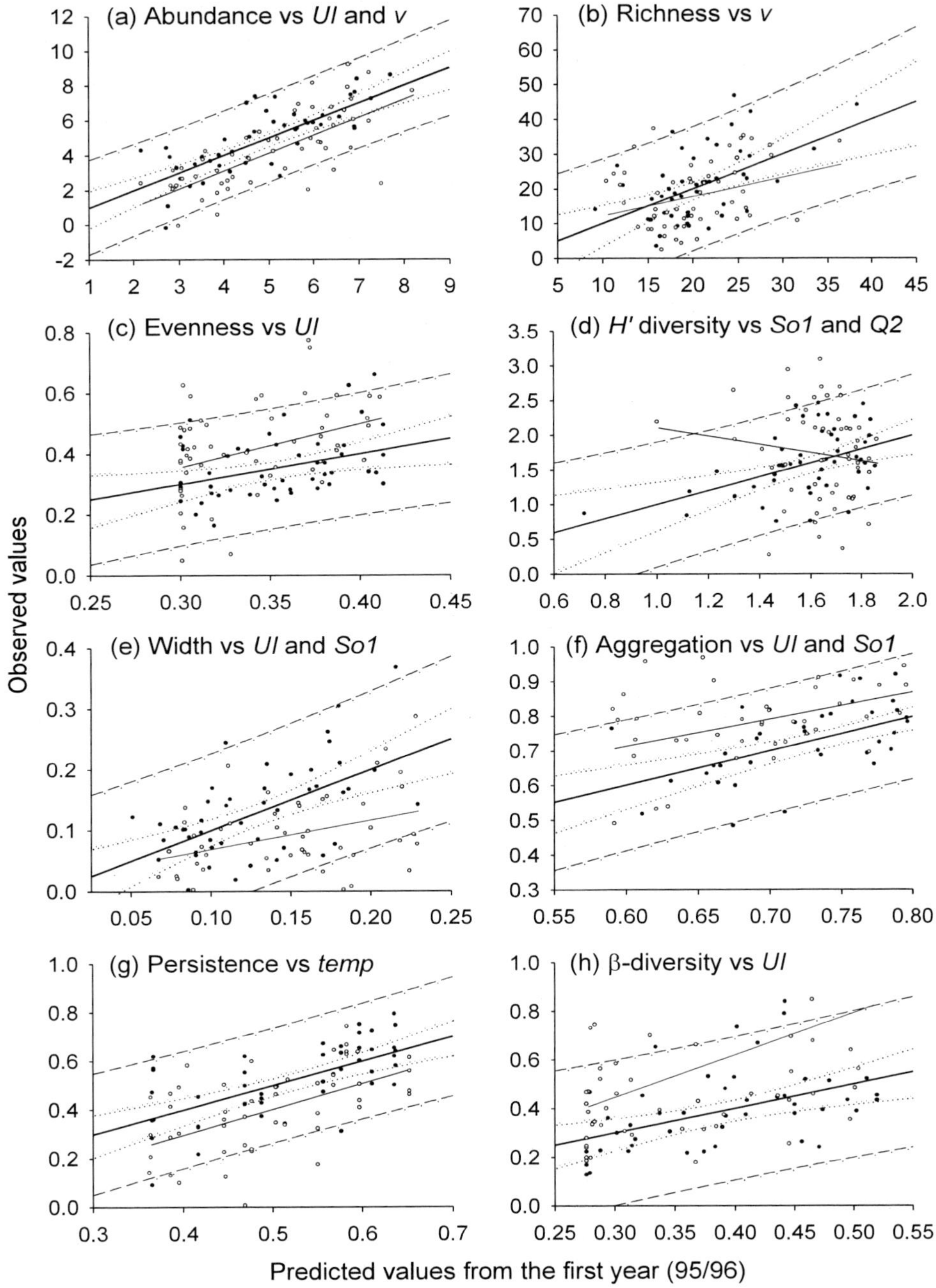

Figure 21. Regressions between observed and predicted values of community attributes estimated from environmental factors. The values of the first year (95/96) are given by the full circles, the corresponding regression by the bold lines (with intercept = 0 and slope = 1) together with the 95% confidence interval (dotted lines) and the 95% prediction intervals (dashed lines). Open circles indicate the observed against predicted values of the subsequent years (96/97 and 97/98). The corresponding regression is shown by the solid thinner line. The coefficients of determination are given in Table 10a (columns r^2 95/96 and r^2 *pred* / 96-98).

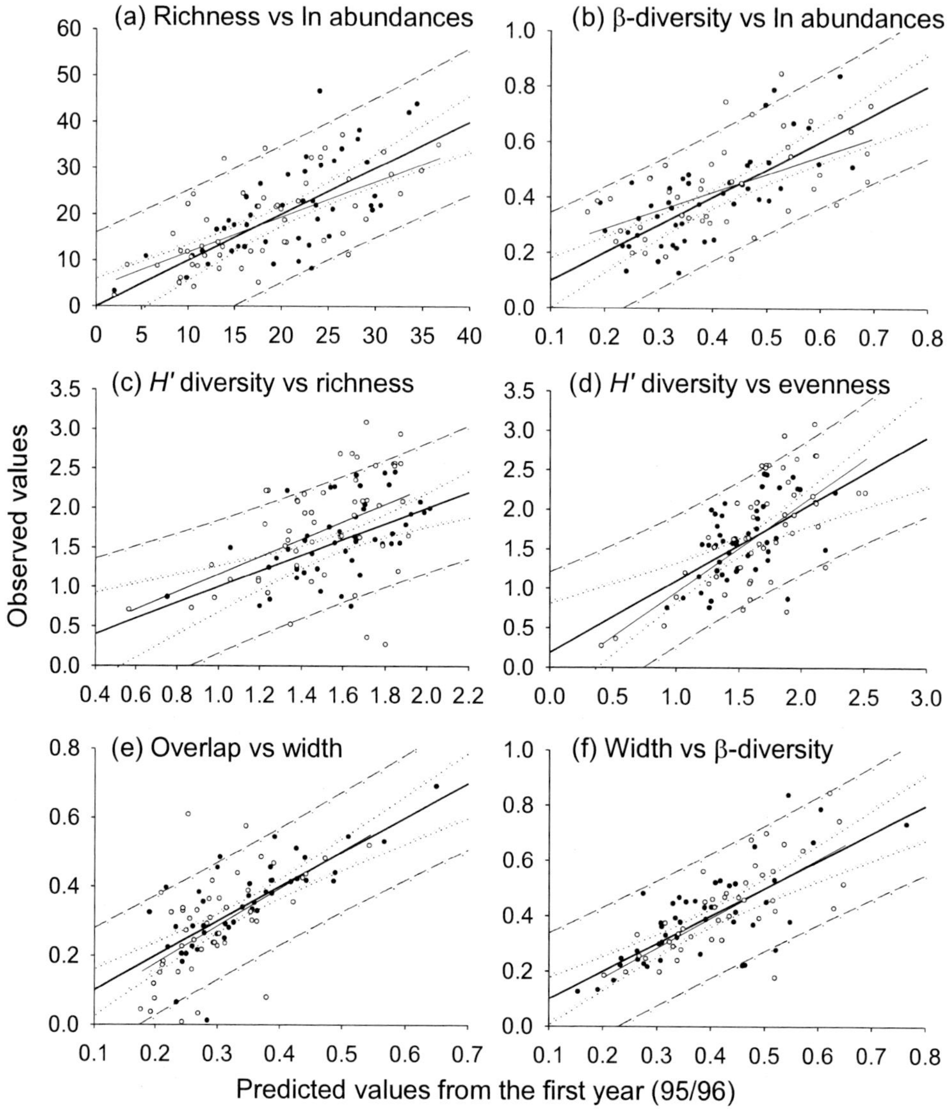

Figure 22. Regressions between observed and predicted values of relationships among community attributes (see legend to Figure 20 for further explanations). The coefficients of determination are given in Table 10b (columns r^2 95/96 and r^2 *pred* / 96-98).

The environmental variables are, as mentioned earlier, sediment turnover (the reverse of habitat stability), water velocity, heterogeneity of the sediment, median of the grain distribution, pore space and water temperature. They differed significantly between sites (see Table 6) but are similar when comparing the means of the three years. Total numbers of individuals, species richness, evenness, diversity are used to characterise the structure of the chironomid assemblages; mean resource width and overlap as well as the degree of spatial aggregation of the different species describe the spatial resource utilisation. Temporal and spatial turnover of the species are measured by compositional differences between successive sampling dates and beta-diversity, respectively.

Habitat stability and water velocity are the strongest predictors for total abundances (52% of the variance in observed values explained by the predicted values; Figure 21a), followed by community persistence (30%, predicted from water temperature; Figure 21g) and species aggregation (25%, habitat stability and heterogeneity; Figure 21f). Evenness (Figure 21c), mean spatial resource width (Figure 21e), and beta-diversity (Figure 21h) account for 14% predictability each, with again habitat stability as the dominant explanatory variable (see also Table 10a).

Table 10. Significant differences between regression coefficients of observed community attribute-values from the years 1996/97, 97/98 and 96–98 combined, and predicted values from the first year (95/96)

(a) Community attributes estimated from environmental parameters

		r^2	r^2 *pred*			Slope			Intercept		
Dep.	Indep.	95/96	96/97	97/98	96-98	96/97	97/98	96-98	96/97	97/98	96-98
Abln	*UI, v*	0.57	0.59	0.45	0.52	ns	ns	ns	ns	Ns	ns
Rich	*V*	0.27	ns	0.30	0.10	*	ns	ns	ns	Ns	ns
Even	*UI*	0.12	ns	ns	0.14	ns	ns	ns	ns	Ns	*
Div	*Sol, Q2*	0.22	ns	ns	ns	*	ns	**	*	Ns	**
Width	*UI, Sol*	0.33	0.27	ns	0.14	ns	**	**	ns	***	***
Over	–	–	–	–	–	–	–	–	–	–	–
Agg	*UI, Sol*	0.42	0.43	ns	0.25	ns	ns	ns	ns	Ns	***
Pers	*Temp*	0.43	0.28	0.34	0.30	ns	ns	ns	***	**	***
Beta	*UI*	0.22	0.21	ns	0.14	ns	ns	ns	ns	*	ns

(b) Relationships among community attributes

		r^2	r^2 *pred*			Slope			Intercept		
Dep.	Indep.	95/96	96/97	97/98	96-98	96/97	97/98	96-98	96/97	97/98	96-98
Rich	Abln	0.49	0.35	0.63	0.44	ns	ns	ns	ns	Ns	ns
Beta	Abln	0.52	0.20	0.42	0.33	*	ns	*	***	***	***
Div	Rich	0.30	0.16	0.25	0.23	ns	ns	ns	ns	Ns	ns
Div	Even	0.32	0.46	0.55	0.63	ns	ns	*	***	***	***
Even	Rich	0.16	0.15	ns	0.09	ns	ns	ns	ns	Ns	***
Over	Width	0.58	0.40	0.35	0.36	ns	ns	ns	***	***	***
Width	Agg	0.63	0.80	0.70	0.75	***	***	***	***	***	***
Over	Agg	0.41	0.38	0.17	0.28	ns	ns	ns	***	Ns	***
Beta	Width	0.55	0.62	0.46	0.55	ns	ns	ns	**	Ns	***

Slopes are tested against 1 and intercepts are tested against 0. Dep = dependent variable; indep = independent variable(s); ns = not significant; * = P < 0.05; ** = P < 0.01; *** = P < 0.001; r^2 95/96 = coefficient of determination from first year; r^2 *pred* = coefficients of determination between observed values from the respective years against predicted values. Abln = ln abundances (individuals per m^2); Rich = species richness; Even = evenness; Div = *H'* diversity; Width = resource width; Over = resource overlap; Agg = spatial aggregation; Pers = temporal persistence; Beta = beta-diversity. Further abbreviations see legend to Table 6

In contrast, predictability is much higher among community attributes, resource width 75%, predicted from spatial aggregation; Table 22b) followed by *H'* diversity (63%, predicted from eveness; Figure 22d) and beta-diversity (55%, mean spatial resource width; Figure 22f). Richness (44%, abundance; Figure 22a) and resource overlap (36%, resource width; Figure 22e; see also Table 10b). Since many slopes of the relationships are not significantly different between the years, deviations from the predictions are mainly due to differences in absolute values of community attributes.

Multivariate patterns of the relationships between community attributes and environmental variables are also analysed by using Redundancy analyses (RDA) for each year separately and all years combined (Figure 23).

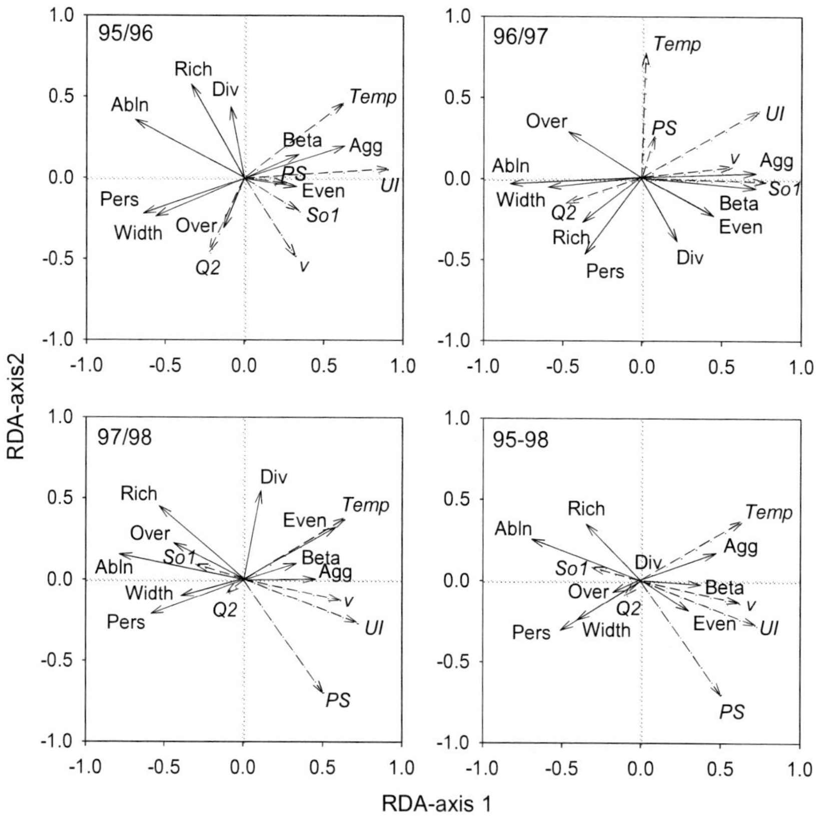

Figure 23. Biplots of Redundancy analyses (RDA) for the single years (95/96, 96/97, 97/98) and all years combined (95-98). Environmental factors are indicated by dashed arrows, community attributes are given by solid arrows. PS, pore space. All other abbreviations are given in Table 9.

The first two axes extracted from the Redundancy analyses explain 81% (first year), 83% (second year), 86% (third year) and 86% (all years combined) of the variance in the attribute–environment relationships. In all three years, habitat stability is the most significant one of the variables for the ordination. Increasing sediment turnover (UI) but water velocity as well, coincide with a decline in number of individuals (Abln), richness (Rich), resource width (Width) and persistence (Pers) whereas evenness (Even), spatial aggregation (Agg) and beta-diversity (Beta) increased. Comparisons of these relationships, by means of Procrustes analyses, show no significant differences between the years (Figure 23). However, univariate and multivariate tests suggest a marked annual change in community composition, mainly with respect to patterns in dominance structure due to the replacement of the most abundant species, in particular, *Cladotanytarsus* cf. *vanderwulpi*, *C.* cf. *daviesi* and *Eukiefferiella* cf. *lobifera* Goethegbuer which were most abundant during the first year, markedly decreased during the subsequent years. In contrast, *Rheotanytarsus rhenanus* Klink had low abundances during the first year but was by far the most dominant species during the third year. The dominance structure changed considerably among years.

On the basis of discriminant functions obtained from environmental factors in the first year, 90% of all samples from the two subsequent years were correctly classified to the four sampling sites, and correct classification on the basis of community attributes accounted for 52% where evenness provided the highest discriminatory significance. It is concluded that most relationships, especially those among community attributes, were consistent for the three years. Because shifts in abundant species were reflected neither by total abundances nor by species richness. Therefore estimation of dominance structure has to be an essential tool for management purposes.

Long-term investigations like the present one provide valuable information on the base-line variation of natural communities in terms of a data–basis for testable ecological theories and management strategies. From these results it seems that for management or conservation purposes, it is not sufficient to rely only upon estimates of abundance and species richness, because they alone are not able to reflect the condition of the assemblages, at least at the local scale. It has to be emphasized that both variables did not change significantly between the years, and changes in dominance structure may easily be overlooked. Furthermore, the decline of even dominant species and their replacement by another was not obvious from abundance and the number of species, and differences in dominance structure render predictions of community structure rather difficult. However, the best predictions were for total abundance, which was in turn a good predictor for species richness and consequently for spatial patterns.

Conclusion

The comparison of different datasets is difficult, due to different sampling methods and standards. A comparison between different sampling techniques for a large, fast flowing river was provided by Bretschko & Schönbauer (1998). The herein presented material from the free-flowing section of the Volga was collected with a bottom grab (Schletterer & Füreder 2010), respectively a modified Multi-habitat-sampling method using a kick-net at the monitoring sites (see Schletterer et al. 2011), for both methods mesh size was 500 µm. From the free-flowing section of the Danube below Vienna, material was collected with a modified Petersen grab (Humpesch & Elliott 1990), mesh size was 200 µm. During the JDS 2 samples were taken using the Airlift technique (Pehofer 1998), in the tributaries multicorer techniques as well as modified multi-habitat-sampling (see Graf et al. 2008a) were used and the mesh size was 500 µm. Despite of difficulties due to different datasets, within this chapter we described general characterisics based on the fauna of the Volga and Danube respectively. We provided a view on the nowadays endangered fauna of the potamon, characterizing also specifics of the investigated river systems. A zoogeographical account will be provided in Graf & Schletterer (in prep). Data from the free-flowing section of the Volga (between the Upper Volga Lakes and Tver) and Danube (below Vienna) was similar in terms of abundances (ind. m^{-2}), during the JDS 2 slightly higher abundances were obtained. However, there is still a restricted insight on the overall conditions at the bottom of large rivers in spatial and temporal terms, thus this should be a focus of future research.

It seems that in the Volga River other neozoan species than in the Danube occur (see e.g. de Vaate et al. 2002). The Danube as an ancient human waterway must be regarded as heavily impacted since at least 200 years. The weakened indigenous fauna – by habitat degradation due to chanalisation, damming, pollution and neozoa – is nowadays a very rare segment or even has already disappeared. A comparison between the Volga and the Danube showed obviously that the headwaters of the Volga River are not tangled by the stressor "neozoa" in the same degree as the Upper Danube. The Danube River is inhabited by more neozoa in respect of species numbers and densities and the indigenous fauna is seriously altered. We are not able to change the ongoing processes, but a detailed monitoring program should document successional developments. In contrast the free-flowing section the Volga River still holds typical elements of the potamon and is therefore a valuable refuge on European Scale (Schletterer & Füreder 2009, 2010a). Similar, the Lafnitz-Rába system (Austria, Hungary) acts as a refugium for potamal species in Europe (Graf & Kovács 2002), but the sensibility of linear systems like rivers is in densely populated areas extremely high. This is underlined by the recent accident in Hungary (Ajka toxic spill; see ORF 2010), which could degrade potamal habitats forever. Consequently a protection of the mentioned hot spots for potamal fauna in Central and East European freshwater ecosystem is absolutely needed, because they hold many species that became rare or even extinct in Europe.

Considering biodiversity, i.e. species distribution patters, the question on the particulate function of different species within the community rises. The functional role of many species, especially of rare species, is still unknown and has led to the theory of redundant species, which might have similar position within ecosystems. But such functionally similar species may diverge in their importance in different states of the ecosystem. Instead of focusing on the conservation of single species, it might be more promising to conserve the habitats where they live, in order to conserve the potamal fauna. Although the relationship between biodiversity and ecosystem functioning and stability is still unclear, there are three complementary arguments why the conservation of biodiversity is important:

- The moral imperative "we are only temporary stewards of planet earth and have a duty to hand it on to future generations in the same state or nearly in the same state, as that in which we found it".
- The balance of nature "there are predictable and unpredictable consequences of loss of biodiversity and some of these consequences might be severely disadvantageous to man - the removal of *keystone* species can alter the whole structure of a natural community."
- The natural resource argument "plants and animals contain exploitable genetic resources that can be captured or modified by man for use in agriculture, industry and medicine."

The indigenous fauna of large rivers is highly endangered. Due to habitat degradation and fragmentation, invasive species and competition phenomena, recolonisation from refugial areas can be expected in very few cases. Species like e.g. *Palingenia longicauda, Prosopistoma pennigerum* (Ephemeroptera), *Agnetina elegantula, Marthamea vitripennis* (Plecoptera), *Stactobiella risi, Platyphylax frauenfeldi* (Trichoptera) or *Macronychus quadrimaculatus* and *Normandia* sp. (Coleoptera) live nowadays in tiny populations in discrete refugia. Biodiversity seems to decline steadily while euryoecious species take over. As rivers are longitudinal habitats each single accident within the catchment can erase populations of European-wide importance for ever. An increase of awareness in combination with an implementation of these flagship species in international protection standards is necessary to assure both a survival of these ancient elements and a functional ecosystem.

ACKNOWLEDGMENTS

Thanks to Dr. L. Füreder, Dr. H. Pehofer (both: University of Innsbruck) and DI Patrick Leitner (BOKU) for discussions and advises. We also appreciate the comments of Dr. M.A. Beketov as well as the suggestions of an anonymous reviewer, which improved the manuscript. The first author was supported by a schorship granted by the Rotary Club Innsbruck ("Forschungsstipendium des Rotary Club Innsbruck").

REFERENCES

Baer, K.E. (1855). Kaspische Studien 1. In: *Bulletin de la Classe physico-mathématique de l'Académie Impériale des Sciences de St.-Pétersbourg,* 13, 193-210.

Bakanov, A.I. (2003). Present-Day State of Zoobenthos in the Upper Volga Reservoirs. *Water Resources, 3*0, 559-568.

Banning, M., Leuchs, H., Rutten, M., Schleuter, M., Scholl, F., & Tittizer, T. (1990): *Die Bundeswasserstrasen als Lebensraum fur Tiere.* – Sonderdruck, Bundesanstalt f. Gewasserkunde, 25 p.

Bäthe, J. (1997). Über die Wiederbesiedlung der Weser durch Ephoron virgo (Olivier 1791) (Ephemeroptera, Polymitarcidae), Anodonta anatina (Linnaeus 1758) und Unio pictorum (Linnaeus 1758) (Lamellibranchiata, Unionidae). *Lauterbornia,* 28, 45–50.

Behning, A. (1924). Zur Erforschung der am Flussboden der Wolga lebenden Organismen. *Monogr. d. Biolog.* Wolga-Station 1 (in Russian).

Behning, A. (1928). *Das Leben der Wolga. Zugleich eine Einführung in die Fluss-Biologie.* Schweizerbart.

Beketov, M.A. & Liess, M. (2008). An indicator for effects of organic toxicants on lotic invertebrate communities: Independence of confounding environmental factors over an extensive river continuum. *Environmental Pollution,* 156, 980-987.

Beketov, M.A., Foit, K., Schäfer, R.B., Schriever, C.A., Sacchi, A., Capri, E., Biggs, J., Wells, C., & Liess, M. (2009). SPEAR indicates pesticide effects in streams – comparative use of speciesand family-level biomonitoring data. *Environmental Pollution,* 157, 1841-1848.

Bernerth, H. & Stein, S. (2003). *Crangonyx pseudogracilis* und *Corophium robustum* (Amphipoda), zwei neue Einwanderer im hessischen Main sowie Erstnachweis für Deutschland von *C. robustum. Lauterbornia,* 48, 45-57.

Berthold, E. & Kaiser, I. (2005). Weitere Funde von *Crangonyx pseudogracilis* und *Chelicocorophium robustum* (Amphipoda) im Main. *Lauterbornia*, 50, 15-17.

Botosaneau, L., & Mlitzky, H. (1978). *Trichoptera.* In: Illies J. (Ed.): Limnofauna Europaea. Eine Zusammenstellung aller die europäischen Binnengewässer bewohnenden mehrzelligen Tierarten mit Angaben über ihre Verbreitung und Ökologie, pp. 333-359.

Bretschko, G., & Schönbauer, B. (1998). Quantitative sampling of benthic fauna in a large river, fast flowing river (Austrian Danube). *Archiv für Hydrobiologie, Supplement Large Rivers*, 11, 195-211.

Brezina, H. (1906). Die Donau vom Leopoldsberg bis Preßburg, die Abwässer der Stadt Wien und deren Schicksal nach ihrer Einmündung in den Strom. *Zeitschrift für Hygiene,* 53, 112-134.

Brunnthaler, J. (1900). Das Phytoplankton des Donaustromes bei Wien. *Verhandlungen der Zoologisch-Botanischen Gesellschaft in Wien,* 50, 308-327.

Butorin, N.V., & Mordukhai-Boltovskoi, F.D. (Ed.) (1978). *Volga i ee Zishn* ("The Volga River and its life"). Academy of Science of the SSSR; Leningrad, Publisher: Nauka (in Russian).

BSE (1951). *Big Soviet Encyclopedia* Vol.8, pp. 602-612.

CGI - Congrès Géologique International (1971). *Carte Géologique Internationale de l' Europe et des régiones riveraines de la Méditerranee (1 : 5.000.000),* Publiée par la Bundesanstalt für Bodenforschung et l'UNESCO. Hamburg.

Damgaard, J. (2005). Distribution, phenology and conservation status of three rare water bugs: Aquarius najas (De Geer, 1773), Aphelocheirus aestivalis (Fabricius, 1794) and Sigara hellensii (C.R. Sahlberg, 1819) from lotic waters in Denmark (Insecta, Hemiptera-Heteroptera: Gerromorpha and Nepomorpha). *Entomologiske Meddelelser*, 73, 25-38.

Dolédec, S., & Statzner, B. (2008). Invertebrate traits for the biomonitoring of large European rivers: an assessment of specific types of human impact. Freshwater Biology, 53, 617–634.

Dolginow, J., & Kropatschjow, S. (1994). *Abriß der Geologie Russlands und der angrenzenden Staaten.* Schweizerbart, Stuttgart.

Dynesius, M., & Nilsson, C. (1994). Fragmentation and Flow Regulation of River Systems in the Northern Third of the World. *Science,* 266, 753-762.

EC (1992): Council Directive 92/43/EEC of 21 *May 1992 on the conservation of natural habitats and of wild fauna and flora,* http://eur-lex.europa.eu/

Eggers, T.O. (2006). Auswirkungen anthropogener Strukturen auf die Makrozoobenthoszönose von Schifffahrtsstraßen - Vergleich einer freifließenden Wasserstraße (Mittlere Elbe) mit einem Schifffahrtskanal (Mittellandkanal) und ihre Bedeutung für Neozoen. - *Dissertation*, Fakultät für Lebenswissenschaften, Technische Universität Carolo-Wilhelmina, 188 pp., Braunschweig.

Fesl, C. (2002). Biodiversity and resource use of larval chironomids in relation to environmental factors in a large river. *Freshwater Biology,* 47, 1065-1087.

Fesl, C., & Humpesch, U.H. (2003). Community structure and resource use of oligochaetes (Annelida) in relation to hydrophysical factors in a large river. Archiv für Hydrobiologie, Supplement Volume 147, *Large Rivers* 14, 307-326.

Fesl, C., & Humpesch, U.H. (2006a). Spatio-temporal variability of benthic macroinvertebrate community attributes and their relationship to environmental factors in a large river (Danube,Austria). *Archiv für Hydrobiologie, Supplement Volume* 158, Large Rivers 16, 329 -350.

Fesl, C., & Humpesch, U.H. (2006b). Community organization of larval chironomid (Diptera) assemblages and environmental factors in a large river: prediction and validation of their interrelationships after three years. *Archiv für Hydrobiologie, Supplement Volume* 158, *Large Rivers,* 16, 495-523.

Fesl, C., Humpesch, U.H., & Aschauer, A. (1999). The relationship between habitat structure and biodiversity of the macrozoobenthos in the free-flowing section of the Danube in Austria – east of Vienna (preliminary results). *Archiv für Hydrobiologie, Supplement Volume* 115, 349-374.

Fesl, C., Humpesch, U.H., & Wöss, E.R. (2005). Biodiversität des Makrozoobenthos der österreichischen Donau unter Berücksichtigung quantitativer befunde der Freien Fließstrecke unterhalb Wiens. *Densia,* 16, 139-158.

Filinova, E.I., Malinina, Y.A., & Shlyakhtin, G.V. (2008). Bioinvasions in Macrozoobenthos of the Volgograd Reservoir. *Russian Journal of Ecology*, 39, 193-197.

Fittkau, E.J., & Reiss, F. (1983). Versuch einer Rekonstruktion der Fauna europäischer Ströme und ihrer Auen. *Arch. Hydrobiol.,* Stuttgart, 97, 1-6.

Fochetti, R., & Tierno de Figueroa, J.M. (2006). Notes on diversity and conservation of the European fauna of Plecoptera (Insecta). *Journal of Natural History*, 40, 2361-2369.

Fochetti, R., & Tierno de Figueroa, J.M. (2008). Global diversity of stoneflies (Plecoptera; Insecta) in freshwater. *Hydrobiol.,* 595, 365-377.

Galewski, K., & Tranda, E. (1978). *Chrząszcze (Coleoptera), Rodziny Pływakowate (Dytiscidae), Flisakowate (Haliplidae), Mokrzelicowate (Hygrobiidae), Krętakowate (Gyrinidae). Fauna słodkowodna Polski. Volume 10.* [Beetles (Coleoptera), the Families of Diving Beetles (Dytiscidae), Crawling Water Beetles (Haliplidae), Squeak Beetles (Hygrobiidae), and Whirligig Beetles (Gyrinidae). The Freshwater Fauna of Poland]. PWN, Warszawa, Poznań, 396 pp (in Polish).

Galil, B.S., Nehring, S., & Panov, V.E. (2007). Waterways as invasion highways – Impact of climate change and globalization. In: Nentwig W: Biological Invasions. *Ecological Studies,* 193, 59–74.

Geest, H.G. van der, Greve, G.D., Kroon, A., Kuijl, S., Kraak, M.H.S., & Admiraal, W. (2000). Sensitivity of characteristic riverine insects, the caddisfly Cyrnus trimaculatus and the mayfly Ephoron virgo, to copper and diazinon. *Environmental Pollution,* 109, 177-182.

Gorski, K., Winter, H.V., de Leeuw, J.J., Minin, A.E., & Nagelkerke, A.J. (2010). Fish spawning in a large temperate floodplain: the role of flooding and temperature. *Freshwater Biology,* 55, 1509-1519.

Graf, W. (1999). Checkliste der Steinfliegen (Insecta: Plecoptera) *Österreichs. Lauterbornia,* 37, 35-47.

Graf, W., & Kovács, T. (2002). The aquatic invertebrates oft he Lafnitz-Rába river system in Austria and Hungary – a natural heritage of the Central European potamocoen. *Internat. Assoc. Danube Res.*, 34, 295-303.

Graf, W., Csányi, B., Leitner, B., Paunovic, M., Chiriac, G., Stubauer, I., Ofenböck, T., & Wagner, F. (2008a). *Macroinvertebrates.* In: Liška, I., Wagner, F., & Slobodník, J.

(Eds.): *Joint Danube Survey.* Final Scientific Report., 41-53; ICPDR – International Commission for the Protection of the Danube River, Wien.

Graf, W., Murphy, J., Dahl, J., Zamora-Munoz, C., & Lopez-Rodriguez, M.J. (2008b). *Distribution and Ecological Preferences of European Freshwater Organisms Volume 1.* Trichoptera. Pensoft Publishers (Sofia-Moscow) 388pp.

Graf, W., Lorenz, A.W., Tierno de Figueroa, J.M., Lücke, S., López-Rodríguez, M.J., & Davies, C. (2009). *Distribution and Ecological Preferences of European Freshwater Organisms (Series editors: Astrid Schmidt-Kloiber and Daniel Hering).* Volume 2. Plecoptera. 262pp.

Graf W. (2010). Eine aktualisierte Checkliste der Steinfliegen (Insecta: Plecoptera) Österreichs. *Lauterbornia* 71: 175-183.

Graf, W., & Schletterer, M. (in prep.). *Zoogeographical aspects of the benthic fauna of Central and East European rivers.*

Grohs, H. (1943). Limnologische Untersuchungen zweier Donaualtwässer bei Wien. *Archiv für Hydrobiologie,* 39, 369-402.

Grosina, H. (1985). Vorstudie für das Forschungsobjekt "Ökosystemstudie Donaustau". *Universitätsverlag Wagner,* Innsbruck.

Gruszka, P. (1999). The River Odra Estuary as a Gateway for Alien Species Immigration to the Baltic Sea Basin. *Acta hydrochimica et hydrobiologica,* 27, 374-382.

Gugel J. (2000). High abundance of Trochospongilla horrida (Porifera, Spongillidae) in the Rhine (Germany) 1992–1995. *Hydrobiologia,* 421, 199–207.

Gurtz, J., Vitvar, T., & Zappa, M. (1999). The response of the water flows of the boreal forest region at the Volga's source area to climatic and land-use changes: *Volga Forest.* 20 p., First year progress report – detailed version: EU-Project INCO – COPERNICUS (Contract n° ERBIC 15-CT98-0120, BBW n° 98.0005).

Hary, N., & Janauer, G.A. (1988). *Interdisziplinäre Studie Donau. Planung + Umwelt* 3, 42-58.

Hary, N., & Nachtnebel, H.-P. (eds.) (1989). *Ökosystemstudie Donaustau Altenwörth.* Universitätsverlag Wagner, Innsbruck.

Humpesch, U.H. (1992). Ecosystem study Altenwörth: impacts of a hydroelectric power - station on the River Danube in Austria. *Freshwater Forum,* 1, 33-58.

Humpesch, U.H. (1996). Case study – the River Danube in Austria. Archiv für Hydrobiologie, Supplement Volume 113, *Large Rivers* 10, 239-266.

Humpesch, U.H., & Elliott, J.M. (eds.) (1990). Methods of biological sampling in a large deep river - the Danube in Austria. *Wasser und Abwasser,* Supplement 2, 83 pp.

Humpesch, U.H., & Fesl, C. (2005). Biodiversity of macrozoobenthos in a large river, the Austrian Danube, including quantitative studies in a free-flowing stretch below Vienna: a short review. *Freshwater Forum,* 24, 3 – 23.

Harrison, S.S., Ross, J. & Lawton, J.H. (1992). Beta diversity on geographic gradients in Britain. *Journal of Animal Ecology,* 61, 151-158.

Hurlbert, S.H. (1978). The measurement of niche overlap and some relatives. *Ecology,* 59, 67-77.

Illies, J., & Botosaneanu, L. (1963). Problèmes et méthodes de la classification et de la zonation écologique des aux courantes, considérées surtout du pont de vue faunistique. *Verhandlungen der internationalen Vereinigung für theoretische und angewandte Limnologie,* 12, 1-57.

Illies, J. (1978). *Limnofauna Europaea*. – Fischer Verlag.

Janauer, G. A. (1996). Aquatic - terrestrial ecotones: the ÖDM-project. Archiv für Hydrobiologie. Supplement Volume 113, *Large Rivers* 10, 211-218.

Janauer, G.A., & Hary, N. (eds.) (2003). *Ökotone Donau-March*. - Universitätsverlag Wagner, Innsbruck.

Jueg, U., & Zettler, M.L. (2004). *Die Molluskenfauna der Elbe in Mecklenburg-Vorpommern mit Erstnachweis der Grobgerippten Körbchenmuschel Corbicula fluminea* (O.F. Müller 1756). Mitteilungen der Naturforschenden Gesellschaft West-Mecklenburg, 4, 83-87.

Karpunin, A.M., Mamonov, S.V., Mironenko, O.A., & Sokolov, A.R. (1997). *Geologicheskiye pamjatniki prirody Rossii.* Lorien, Moscow

Keffermüller, M. (1957). W sprawie ochrony pewnych odcinków środkowego biegu Warty. *Przyroda Polski Zachodniej*, 1, 152-155.

Keffermüller, M. (1959). Nowe dane dotyczące jętek (Ephemeroptera) z rodzaju Ametropus Alb. i Behningia Lest. *Pozn. Tow. Przyj. Nauk, Prace komisji biologicznej,* 19, 1-32.

Keffermüller, M. (1963). Masowy lot Beningii (Behningia lestagei Motas et Bacesco, Ephemeroptera) w Ostrowsku nad Wartą. *Bad. Fizjogr. Pol. Zach.*, 12, 323-325.

Kłonowska-Olejnik, M. (2009). *Polska czerwonka księga zwierząt – bezkręgowce – Behningia ulmeri Lestage* 1929, www.iop.krakow.pl

Kessler, K.F. (1877). *Riby, vodyashchiesya i vstrechayushchiesya v Aralo-kaspiisko-pontiiskoi ikhtiologicheskoi moblasti* (Fishes of the Aralo-Caspian-Pontine region). Trudy Aralo-Kaspiiskoi ekspeditsii, 4, 190–196 (in Russian).

Khaiter, P. (1995). Hydrobiological responses to eutrophication: Volga Reservoirs, Russia. IAHS Publication 230 *„Man's Influence on Freshwater Ecosystems and Water Use":* 185-190.

Kheraskova, T.N., Andreeva, N.K., Vorontsov, A.K., & Kagramanyan, N.A. (2005). Evolution of the Moscow Sedimentary Basin in the Early Paleozoic. *Lithology and Mineral Resources,* Vol. 40, No. 2, pp. 150–166.

Kopylov, A.I. (Ed.) (2001). Ecological problems of the Upper Volga, Yaroslavl, pp. 1-427.

Kureck, A. (1992). Das Massenschwarmen der Eintagsfliegen am Rhein – Zur Rückkehr von *Ephoron virgo* (Olivier 1781). *Natur und Landschaft,* 67, 407–409.

Kuzovlev, V.V., & Schletterer, M. (2006). Upper Volga Expedition 2005 - Technical Report. – *Proceedings of Freshwater Research,* 1, 1-140.

Lepneva, S.G. (1970). Fauna of the USSR. Trichoptera, Vol. 2 Nr. 1: *Larvae and pupae of the Annulipalpia*. Isreal programm of scientific translations.

Lillehammer, A. (1988). *Stoneflies (Plecoptera) of Fennoscandia and Denmark.* Fauna entomologica Scandinavica 21.

Liepolt, R. (1961). *Limnologische Forschungen im österreichischen Donaustrom. Verhandlungen der internationalen Vereinigung für theoretische und angewandte Limnologie,* XIV, 422-429.

Liepolt, R. (ed.) (1965, 1966). Limnologie der Donau; *IAD-Monograph I and II.* E. Schweizerbart, Stuttgart.

Liess, M., Schäfer, R.B., & Schriever, C.A. (2008). The footprint of pesticide stress in communities – species traits reveal community effects of toxicants. *Science of the Total Environment,* 406, 484-490.

Lawton, J.H., & May, R.M. (eds.) (1995). *Extinction rates*. Oxford University Press.

Literáthy, P., Koller-Kreimel, V., & Liska, I. (eds.) (2002). *Joint Danube Survey.- Technical Report of the International Commission for the Protection of the Danube River*, 261 pp.

Litvinov, A.S., Mineeva, N.M., Papchenkov, V.G., Korneva, L.G., Lazareva, V.A., Shcherbina, G.K., Gerasimov, Y.V., Dvinskikh, S.A., Noskov, V.M., Kitaev, A.B., Alexevnina, M.S., Presnova, E.V., Seletkova, E.B., Zinovfev, E.A., Baklanov, M.A., Okhapkin, A.G., & Shurganova, G.V. (2009). Volga River Basin, pp. 23-57, in: Tockner, C., Robinson, C.T., & Uehlinger, U. (Ed.): *Rivers of Europe*, Academic Press, 728 p.

Lövei, G.L. (1997). Global change through invasions. *Nature,* 388, 627-628.

Manley, B.F.J. (1997). *Randomization, bootstrap and Monte Carlo methods in biology.* Chapman & Hall, London.

Middelkoop, H., Schoor, M.M., & Babich, D.B. (2005). Bio-morphodynamics of the Lower Volga river: a reference for river rehabilitation in The Netherlands. *Archiv fuer Hydrobiologie* (Large Rivers, 15, 89-103.

Milashevich, K.O. (1881). Etudes sur la faune des mollusques vivant terrestres et fluviatiles de Moscou. *Bull. De la Soc. Nat. de Moscou,* 56, 78-91.

Moog, O., Brunner, S., Humpesch, U.H., & Schmidt-Kloiber, A. (2000). The distribution of benthic invertebrates along the Austrian stretch of the River Danube and its relevance as an indicator of zoogeographical and water quality patterns – part 2. *Archiv für Hydrobiologie,* Supplement Volume 115, Large Rivers 11, 473-509.

Morisita, M. (1962). Id-index, a measure of dispersion of individuals. *Researches on Population Ecology,* 4, 1-7.

Motas, C., & Bacesco, M. (1937). La découverte en Roumanie d'une nymphe d'éphemere appartenant au genre Behningia Lestage 1929. *Ann. Sci. de l'Univ. de Jassy*, 24, 25-29.

Nazarova, L.B., Semenov, V.F., Sabirov, R.M., & Efimov, I.Y. (2004). The State of Benthic Communities and Water Quality Evaluation in the Cheboksary Reservoir. *Water Resources,* 31, 316-322.

Oberzill, W. (1941). Biologisch-chemische Untersuchung des Tritonwassers im Gebiete der Alten Donau bei Wien. *Archiv für Hydrobiolologie,* 37, 533-577.

Obrdlik, P., Adamek, Z., & Zahradka, J. (1979). Mayfly fauna (Ephemeroptera) and the biology of species Potomanthus luteus (L.) in a warmed stretch of the Oslava river. *Hydrobiologia,* 67, 129–140.

Ofenböck, T., Moog, O., Hartmann, A., & Stubauer, I., (2007). Fliessgewässer-Qualitats-element Makrozoobenthos: Felderhebung, Probennahme, Probenaufbereitung und Ergebnisermittlung (Leitfaden zur Erhebung der biologischen Qualitätselemente Teil A2. Version A2-01c MZB), Bundesministerium fur Land- und Forstwirtschaft, Umwelt und Wasserwirtschaft (BMLFUW), 214 pp.

ORF (2010*). Giftige Schlammflut in Ungarn – Mit Gips gegen Giftschlamm – Langzeitfolgen klaum abschätzbar*; http://news.orf.at/stories/2018433/2018432/ (October 2010)

Pallas, P.S. (1771). *"Travels through various provinces of the Russian Empire",* 3 Volumes.

Panov, V., Dgebuadze, Y., Shiganova, T., Filippov, A., Shestakov, V., & Minchin, D. (2005). A risk assessment of biological invasions: inland waterways of Europe – the northern invasion corridor case study. Abstract of an Oral presentation, page 55; In: Gherardi, F. (Ed.): *INWAT Workshop:* Biological invasions in inland waters, Firenze.

Parz-Gollner, R., & Herzig, A. (2000). *Ökosystemstudie Donaustau Altenwörth Band II.* Universitätsverlag Wagner, Innsbruck.

Pehofer, H.E. (1998). A new quantitative air-lift sampler for collecting invertebrates designed for operation in deep, fast flowing gravelbed rivers. *Arch. Hydrobiol., Suppl. Large River,* 11, 213-232.

Peres-Neto, P.R. (2000). *Procrustes Analysis and PROTEST.* – Toronto.

Perova, S.N. (1998). Structure of macrozoobenthos of the Gorky Reservoir. *Biol. Vnutr. Vod,* 3, 19–28.

Pesta, O. (1928). Berichte zur Limnologie der Alten Donau bei Wien. *Archiv für Hydrobiologie,* 19, 301-317.

Pianka, E.R. (1973). The structure of lizard communities. *Annual Review of Ecology and Systematics,* 4, 53-74.

Pielou, E.C. (1975). Ecological diversity. – *John Wiley and Sons,* New York, 165 pp.

Pitsch, T. (1983). *Die Trichopteren der Fulda, insbesondere ihre Verbreitung im Flußlängslauf.* Diplomarbeit Freie Universität Berlin, 189 p

Rau, A. (1888). *Handbuch für Käfer-Sammler,* 494 pp, Creutz`sche Verlagsbuchhandlung, Magdeburg.

Rossinsky, D.M. (1892). *Materialy k' poznaniyu fauny bezposvonochnykh Moskvy reki (Materials about the Invertebratefauna of River Moskva).* Izvestiya Imperatorskago Obshchestva Lyubitelei Estestvoznaniya, Antropologiy i Etnografiy, Imperatorckom' Moskovskom' Universitet', 67 (= Dnevnik' Zoologicheskago Otdeleniya Obshchestva i Zoologicheskago Muzeya, 6; Trudy Zoologicheskago Otdeleniya Obshch 6: 1-38 (in Russian).

Rolauffs, P., Stubauer, I., Zahradkova, S., Brabec, K., & Moog, O., (2004). Integration of the saprobic system into the European Union Water Framework Directive—case studies in Austria, Germany and Czech Republic. *Hydrobiologia* 516, 285–329.

Rozenberg, G.S., & Vyhristjuk, L.A. (Eds.) (2008). *Kuibyshevskoe vodohranilishe nauchno-informazionnyh spravochnik.* pp. 123, Kassandra.

Rupp, H., Büttner, O., Krüger, F., Kunert, M., Meißner, R., Muhs, K., Witter, B., & Friese, K. (2000). *Wirkung von Hochwasserereignissen auf die Schadstoffbelastung von Auen und kulturwirtschaftlich genutzte Böden im Überschwemmungsbereich von Oka und Elbe.* Unveröffentlichter Abschlußbericht zum BMBF Forschungsvorhaben (FKZ: 02WT9617/0), Falkenberg, pp. 166.

Russev, B.K. (1967). *Das Zoobenthos der Donau.* – In: Liepolt R., Limnologie der Donau, Lieferung 3, 242-271, Stuttgart.

Sabaneev, V.P. (1880). *Spisok nazemnih i presnovodnyh molluskov, vstrechauhsja v Yaroslavskoi guberny* (List of terrestrial and freshwater molluscs in Yaroslavl-Region). Tr. O-va dlja issledovanie Yaroslavskoi gubernij; T1. (in Russian)

Sahagian, D., Pinous, O., Olferiev, A., & Zakharov, V. (1996). Eustatic Curve for the Middle Jurassic–Cretaceous Based on Russian Platform and Siberian stratigraphy: Zonal Resolution. *American Association of Petroleum Geologists Bulletin*, Vol. 80, 1433–1458.

Schallgruber, F. (1944). Das Plankton des Donaustromes bei Wien in qualitativer und quantitativer Hinsicht. *Archiv fur Hydrobiolologie,* 39, 665-689.

Schiemer, F., & Janauer, G.A. (1994). Monitoring rivers and floodplains. Stapfia, 31, 93-107.

Schiemer, F., & Waidbacher, H. (1992). Strategies for conservation of a Danube fish fauna. In Boom, P.J., Calow, P., & Petts, G.E. (Eds.*), River Conservation and Management,* 364-382.

Schiemer, F. & Zalewski, M. (1992). The importance of riparian ecotones for diversity and productivity of riverine fish communities. *Netherlands Journal of Zoology*, 42, 323-335.

Schiemer, F., Zalewski, M., & Thorpe, J.E. (1995). Land/inland water ecotones: intermediate habitats critical for conservation and management. *Hydrobiologia,* 303, 259-264.

Schiller, J. (1926). Der thermische Einfluß und die Wirkung des Eises auf die planktischen Herbstvegetationen in den Altwässem der Donau bei Wien nach regelmäßigen Beobachtungen von Oktober 1918 bis Ende 1925. *Archiv für Protistenkunde,* 56, 1-62.

Schletterer, M. (2006). Biological Assessment of the Upper Volga River. – In: Kuzovlev, V.V. & Schletterer, M. (eds.), *Upper Volga Expedition 2005 – Technical Report. – Proc. Freshwater Res.,* 1, 76–126.

Schletterer, M., & Eggers, T.O. (2006). Evidence of Freshwater Sponges (Porifera: Spongillidae) in the Upper Volga River (Russia). *Ber. nat.-med. Verein Innsbruck*, 93, 73-84.

Schletterer, M., & Kuzovlev, V.V. (2007). First record of *Prosopistoma pennigerum* from the Russian Federation (Ephemeroptera: Prosopistomatidae), *Zoosystematica Rossica*, 16, 169–172.

Schletterer, M., & Füreder, L. (2008). Entomological notes on all described species of the mayfly-family Prosopistomatidae (Insecta: Ephemeroptera), *Berichte des naturwissenschaftlich- medizinischen Vereines Innsbruck,* 95, 65–75.

Schletterer, M., & Füreder, L. (2009). The family Prosopistomatidae (Ephemeroptera): A review on its ecology and distribution, with particular emphasis on the European species Prosopistoma pennigerum Muller 1785. *Aquatic Insects 31* (Suppl. 1): 603–620.

Schletterer, M. (2009). Benthic Invertebrates and Reference Conditions in East European Running Waters - Case study in the headwaters of Volga River (Tver Region, Russia) and adjacent water bodies. *Doctoral dissertation,* University of Innsbruck, 302 pages

Schletterer, M., & Füreder, L. (2010a). The River Volga headwaters: Inventory, biodiversity and conservation. – *River Systems,* 19, 15–32.

Schletterer, M., & Füreder, L. (2010b). Large river mayflies (Insecta: Ephemeroptera) and their distribution in the Upper Volga River: current knowledge and conservation issues. – *River Systems,* 19, 59–73.

Schletterer, M., Füreder, L., Kuzovlev, V.V., & Beketov, M.A. (2010). Testing the coherence of several macroinvertebrate indices and environmental factors in a large lowland river system (Volga River, Russia). *Ecological Indicators,* 10, 1083-1092.

Schletterer, M., Schönhuber, M., & Füreder, L. (2011). Biodiversity of diatoms and macroinvertebrates in an East European lowland river, the Tudovka River (Tver Region, Russia). *Boreal Env. Res.* 16, in press.

Schletterer, M., & Füreder, L. (2011). The Volga River: research history and synthesis, *River Systems (in press).*

Schletterer, M., & Humpesch, U.H. (in prep.). *Zoobenthos diversity in the Volga delta – a transect from Astrakhan to the Caspian Sea.*

Schoenemund, E. (1930). Eintagsfliegen oder Ephemeroptera.– In: Dahl, F. (ed.), *Tierwelt Deutschlands und der angrenzenden Meeresteile nach ihren Merkmalen und ihrer Lebensweise.* – Verlag G. Fischer.

Shaporenko, S.I., Shilkrot, G.S., Tikhomirova, A.V., & Leonov, A.V. (2002). *Ozero Seliger – vtoroj istok Volhi. Priroda 6* (in Russian).

Shcherbina, G.H. (2009). Change of species composition and structural functional characteristics of macrozoobenthos in water systems of north-western Russia due to natural and anthropogenic factors. *Doctoral Dissertation, Avtoreferant,* pp. 49. (in Russian).

Shiklomanov, I.A., 1999: World Water Resources and their use – A joint SHI/UNESCO product. Online database: http://webworld.unesco.org/water/ihp/db/shiklomanov/.

Skal'skaya, I. A. (2008). Nonindigenous invertebrates in the periphyton and benthos of upper Volga reservoirs: A review. *Inland Water Biology,* 1, 163-174 (in Russian).

Smith-Gill, S.J. (1975). Cytophysiological basis of disruptive pigmentary pattern in the leopard frog Rana pipiens. II. Wild type and mutant cell specific patterns. *Journal of Morphology,* 146, 35-54.

Soldan, T., Zahradkova, S., Helešic, J., Dušek, L., & Landa, V. (1998). Distributional and quantitative Patterns of Ephemeroptera and Plecoptera in the Czech Republic: a possibility of detection of long-term environmental changes of aquatic biotopes. – *Folia Fac. sci. nat. Univ. Masaryk.* Brunensis Biol., 98, 1–305.

Sommerfeld, N., Hein, T., Schneider-Jakoby, M., Baumgartner, C., Ostojić, A., Paunović, M., Bloesch, J., Siber, R., & Tockner, K. (2009). The Danube River Basin, pp. 59-112, in: Tockner, C., Robinson, C.T., & Uehlinger, U. (Ed.): *Rivers of Europe*, Academic Press.

Spreitzer, H. (1935). *Eiszeitstudien in Russland: Geomorphologische und quartärgeologische Untersuchungen zwischen Wolga und Oka.* Jahrbuch der Geographischen Gesellschaft zu Hannover für 1934 und 1935.

Starmühlner, F., Kusel-Fetzmann, E., Steiner, H., & Aschenbrenner, L. (1972). Die Pflanzen und Tierwelt der Donau. In F. Starmühlner & F. Ehrendorfer (eds.) *Naturgeschichte Wiens* 2, 531-576.

Statzner, B. (1979). Der obere und untere Schierenseebach (Schleswig-Holstein). *Strukturen und Funktionen in zwei norddeutschen See-Ausfluss-Systemen, unter besonderer Berücksichtigung der Makroinvertebraten.*- Diss. Univ. Kiel, 551 pp.

Statzner, B., Resh, V.H., & Dolédec, S. (1994). Ecology of the Upper Rhone: a test of habitat templet theories. *Freshwater Biology,* 31, 253-554.

Steuer, A. (1901). Die Entomostrakenfauna der Alten Donau bei Wien. *Zoologische Jahrbücher,* 15, 1-16.

Stundl, K. (1943). Die Donau zwischen Passau und Pressburg als Fischwasser. *Fischereizeitung,* 46, 1-3.

Stundl, K. (1944a). Möglichkeiten der Ertragssteigerung in der Donaufischerei. *Fischereizeitung,* 47, 13-17.

Stundl, K. (1944b). Die Donau zwischen Passau und Reichsgrenze als Fischwasser. *Fischereizeitung,* 47, 34-37.

Stundl, K. (1951). Zur Hydrographie und Biologie der österreichischen Donau. *Schweizerische Zeitschrift für Hydrologie,* 13, 36-43.

Surin, Z.S. (Ed.) (1976). Resources on surface waters of the USSR. Vol. 10. *The Upper-Volga area.* (Resursy poverhnostnih vod SSSR, T.10 Verhne-Volgskiy Rayon, Moscow Dept. of Gidrometeoizdat.

Tittizer, T., Fey, D., Sommerhäuser, M., Málnás, K., & Andrikovics, S. (2008). Versuche zur Wiederansiedlung der Eintagsfliegenart Palingenia longicauda (OLIVIER, 1791) in der Lippe. *Lauterbornia,* 63, 57-75.

Ter Braak, C.J.F., & Prentice, I.C. (1988). A theory of gradient analysis. *Advances in Ecological Research,* 18, 271-317.

Tshernova, O.A. (1938). *Novoe semejstvo podenok (Ephemeroptera).* Izd. Akad. Nauk SSSR, Ser. biol., 1, 129-137.

Ulmer, J.A. (1924). Eine merkwürdige Ephemeridennymphe aus der Wolga. *Arb. Biol. Wolga-Station,* 3, 1-8.

Unesco (2004). The Volga Vision – UNESCO`s interdisciplinary Initiative for the sustainable *Development of the Volga*, pp. 1-144

de Vaate, A.B., Jazdzewski, K., Ketelaars, H.A.M., Gollasch, S.,& Van der Velde, G. (2002). Geographical patterns in range extension of Ponto-Caspian macroinvertebrate species in *Europe. Can. J. Fish. Aquat. Sci.,* 59, 1159-1174.

Vannote, R.L., Minshall, G.W., Cummins, K.W., Sedell, J.R., & Cushing, C.E. (1980). The river continuum concept. *Canadian Journal of Fisheries and Aquatic Siences,* 37, 130-137.

Vornatscher, J. (1938). Faunistische Untersuchung des Lusthauswassers im Wiener Prater. *Internationale Revue der gesamten Hydrobiologie und Hydrographie*, 37, 320-363.

Ward, J.V., Tockner, K., Arscott, D.B., & Claret, C. (2002). Riverine landscape diversity. Freshwater Biology, *47, 517–539.*

Wefering, F., & Nikitina, E. (2005). CABRI-Volga - *Cooperation Along a Big River: Institutional coordination among stakeholders for environmental risk management in the Volga basin.* Briefing paper form: www.cabri-volga.org/

Wollenberg, A.L. (1977). Redundancy analysis, an alternative for canonical analysis. *Psychometrika,* 42, 207-219.

Zelinka, M., & Marvan, P. (1961). Zur Präzisierung der biologischen Klassification der Reinheit fliessender Gewässer, *Arch. Hydrobiol.,* 57, 389-407.

Zinchenko, T.D. (1992). Long-term (30 years) dynamics of Chironomidae (Diptera) fauna in the Kuibyshev water reservoir associated with eutrophication processes. *Aquatic Ecology,* 26, 533-542.

Zwick, P. (1984). *Marthamea beraudi* (Navas) and its European congeners (Plecoptera: Perlidae). *Annls Limnol.,* 20, 129-139.

Zwick, P. (1992). Stream habitat fragmentation - a threat to biodiversity. *Biodiversity and Conservation,* 1, 80-97.

In: River Channels: Types, Dynamics and Changes
Editor: Daniel A. Molina, pp. 51-110
ISBN 978-1-61324-148-6

Chapter 2

LARGE ALLUVIAL RIVER CHANNEL PATTERNS AND COMPLEXITY

Aleksey Sidorchuk
Geographical Faculty, Moscow State University, 199899 Moscow, Russia

ABSTRACT

The geometry of an alluvial river channel is controlled by 1) the complexity of the hierarchy of alluvial forms (the number of levels of hierarchy); 2) the height of alluvial forms relative to the channel depth; 3) the width of alluvial forms relative to the channel width; 4) the hierarchic level of stable (vegetated) alluvial forms; 5) river basin climate and vegetation; wash load concentration in the stream; 6) river valley geological structure and texture; 7) the history of the river valley evolution. Interaction between the river flow and channel bed is the main cause of alluvial forms generation. The hierarchy of alluvial forms consists of 3-8 levels. The complexity of the system generally increases with the river width/depth ratio increase and slope decrease. The typical sequence of alluvial forms is: 1) megaripples (ultramicroforms); 2) dunes of the first and second order (microforms); 3) large dunes of the third order and small bars of the first order (mesoforms); 4) large bars (macroforms). These alluvial forms (usually meso- and macroforms), depending on their relative height, river basin climate and vegetation, and wash load concentration in the stream, can be stabilised by the vegetation and determine the channel pattern. River channel can be called 'meandering' when the width (lateral wavelength) of the largest stabilised (vegetated) alluvial feature, which determines overall channel geometry, is close to the channel double width. Meanders are simple when only these largest bars of hierarchical complex are vegetated. During the low water period smaller non-vegetated bars appear above the water level and the river can be characterised as meandering with alternating and braid bars. The river channel can be called braided by islands if the width (lateral wavelength) of the largest stabilised (vegetated) alluvial feature, which determines channel pattern, is less than the channel double width. The larger the relative channel width is and the lower the hierarchical level of the vegetated alluvial form is, the more complicated the braiding is. During the low water the channel can be braided both by islands and non-vegetated bars. The river channel is braided by bars (at low water) or straight (at high water) when all alluvial forms are non-vegetated and overall channel geometry is controlled only by non-fluvial factors or/and old fluvial relief. Non-fluvial control factors (soil texture and rocks, vegetation in the river valley and slopes), as well as old fluvial relief of floodplain and river terraces, increase the complexity of the river channel patterns.

Keywords: channel flow hydraulics, alluvial feature hierarchy, river basin climate and landscape, floodplain vegetation cover, river channel pattern

List of Notations (If not in the Text)

Q	discharge
W_0	channel width
d_0	channel depth
D_0	depth of unperturbed channel
d'	depth perturbation
H_b	river banks height
h	flow surface elevations
h'	perturbations of the flow surface elevations
z	channel bed elevations
u,v,w	longitudinal, lateral, vertical velocity components at a point
u^*,v^*,w^*	longitudinal, lateral and vertical velocity components at the flow surface or near bed
U	longitudinal velocity of unperturbed flow
u',v',w'	perturbations of the longitudinal, lateral and vertical velocity components at the flow surface
u_*	shear velocity
t	time
X, Y, Z	longitudinal, lateral and vertical curvilinear orthogonal coordinates
s_x,s_y	distances along longitudinal and lateral curvilinear orthogonal coordinates
K_x, K_y	curvature of longitudinal and lateral curvilinear orthogonal coordinates
H_x,H_y	scale factors
L_x,L_y	longitudinal and lateral length of the undulations of channel relief
k_x,k_y	longitudinal and lateral wavenumbers for perturbations
$c = c_r + ic_i$	complex celerity
A,B,P,T	functions of lateral coordinate in (1-13)
L_m	channel meander wavelength
Θ	angle of deflection for meander
q_s, q_b	longitudinal and lateral sediment flux
R	sediment concentration at a point
R_b	bed sediment concentration
R_m	depth-averaged sediment concentration
Δ	roughness element height
ω	fall velocity of sediment particles
ε	bed sediments porosity
Fr	Froude number $U / \sqrt{gD_0}$

C_0	Chezy coefficient
κ	von Karman constant
α_i, β_i	kinematic coefficients
γ_0	coefficient in the formula of Rozovskiy
n	exponent in the formula (1-6)
A_i	coefficients in dispersion equation
g	acceleration due to gravity
I_0, H_0	function of Bessel and Struve of zero order

INTRODUCTION

River channel is the negative form of relief, the lowermost band of a river valley, continuously or discontinuously filled by running water. In general, the configuration of river channel is permanently changing through time due to water level seasonal instability and long-term alluvial relief transformation. Therefore channel pattern definition usually reflects only the most general, oft-recurring and characteristic morphologic and dynamic features of a given river. Thus, a well-knowen classification of river patterns by Leopold and Wolman (1957) consists only of three types: straight, meandering and braided channels. These three types are attached to three main water stages. Channels usually look most straight at high water levels. Alluvial meandering channels are best seen at bankfull (or lower) levels, while braided channels are best seen at low water levels. Further development of channel pattern classification led to recognition of multi-channel features at bankfull level (Knighton, 1998), as channels divided to branches by islands and so called anastomosing systems. Such classifications can be called single-level ones, as channel geometry on the plane surface within each class is described only for one of the main water levels. It is obvious that single-level classifications cannot be general, as the same channel can be attributed to different classes at different water stages: straight channel at the high water level can be considered braided at the low water level. The variability of channel geometry within each class is high and multi-factorial, therefore existing classifications of the channel patterns are rather complicated.

Another classification (Nomenclature…, 1966) is not attached to any particular water stage and describes the hierarchy of alluvial features: the smallest bed forms are called ripples, the largest are called bars, with dunes as the bed forms of intermediate size. This classification development resulted in the increasing number of the levels in the hierarchy. Richards (1980) added the class of megaripples – bed forms smaller than dunes, with size related to flow depth. Allen (1978) further developed the hierarchy of dunes. At some level of this hierarchy non-vegetated features, composed of relatively coarse channel alluvium (sand, gravel, boulders) can be covered and stabilised with vegetation and fine floodplain alluvium. Hierarchical classification is more general than single-level classification and includes the latter.

There are two main approaches in alluvial relief hierarchy investigations: hydrodynamic and morphologic. Hydrodynamic approach deals with mechanics of alluvial relief development due to interaction between water flow and movable channel bed. Dynamics of particular alluvial features or a river channel section is usually investigated with the help of theoretical 2D and 3D flow models and some empirical formulas for channel resistance and sediment transport rate calculation. Morphologic approach is much more empirical. Within this approach, the morphology and dynamics of particular alluvial features and their complexes are described in the large flumes or in the natural rivers. Therefore, the problem of discrete alluvial features distinguishing is of major importance. The combination of both approaches led to a synthetic view on the hierarchy of alluvial relief and is used in the following paper. The paper is based mostly on the author's personal theoretical investigations and observations at various rivers with the use of a very limited number of the other specialists' publications. Therefore the literature review is omitted here due to large number of publications on the problem and existence of several summarising monographs (Bridge, 2003, Gregory, 1977, Knighton, 1998, Leopold, 1994, Richards, 2004, Schumm, 1977, 2005).

PART 1

The Theoretical Approach to River Channel Pattern Estimation

Basic principle in alluvial relief analysis is the principle of interaction between water flow and movable channel bed (Velikanov, 1954). After Makkaveev (1955), this interaction has a form of the flow structure 'imprint' on the movable bed with associated feedback. Quasi-periodic changes of the flow velocity (i.e. flow acceleration and deceleration) result in regular changes of the sediment transport rate and, therefore, fields of bed erosion and sediment deposition. Erosion and deposition cause undulations of the channel bed (sand waves) and led to creation of dune-like initial alluvial relief. Channel bed deformation transforms the depth field and resistance distribution in river channel, and velocity field in flow is adapted to these changes. Some of initial undulations grow due to interaction with flow structures and are transformed into alluvial features with finite amplitude wavelength coherent with flow structure. Others are characterised with negative growth rate and do not form any alluvial features. This interaction results in a dynamic equilibrium of self-organising dynamic system 'stream flow – channel bed', and in the predictable morphological and dynamic pattern of alluvial relief hierarchy. Macro-turbulent quasi-periodic wave-like and eddy-like structures are formed in flow, and complicated hierarchical alluvial relief is developed in channel due to interaction in system 'stream flow – channel bed'. The range of wavelengths of the growing undulation is rather broad. Therefore the resulting alluvial features combine into a hierarchical system, where the larger forms are complicated by the smaller ones.

The Initial Flow and Bed Instability Analysis

The analysis of the initial instability of the quasi-periodic structures in flow and on river channel bed gives a basis for description of the alluvial relief hierarchy generation. The most perspective way to describe this hierarchy is to solve numerically the full system of hydrodynamic equations with the equation of sediment mass continuity. In the majority of works with hydrodynamic description of river channel relief generation (Kennedy, 1963; Callander, 1969; Fredsoe, 1978; Mikhinov, 1985, Sidorchuk, 1992, 1996) the equations of hydrodynamics are parameterised, linearized and solved analytically. Velocity field, elevations of flow surface and channel bed are presented as a sum of two components: main flow and perturbation, the behaviour of the perturbations is analysed. The level of simplification of the basic equations varies in different works, with the major influence to the results.

For example, depth-averaged 2D equations of fluid motion and continuity with the equation of bed deformation were used in the works of Callander (1969), Fredsoe (1978) and Mikhinov (1983). The flow resistance was described in terms of Chezy formula. At the stage of depth averaging of 3D equations of momentum and conservation Callander and Fredsoe used static pressure distribution assumption (Saint-Venant approach). Mikhinov took into account the dynamic pressure vertical distribution and curvature of the flow surface (Boussinesq approach); this assumption led to appearance of the additional mode of instability at the region of short waves (megaripples). In all these solutions the 3D helical flow effects were lost.

At the stage of selection of the initial perturbation geometry Fredsoe used moving waves of pre-determined bi-sinusoidal shape. Callander and Mikhinov used more general form of longitudinal moving sinusoidal waves, and their shape in lateral direction was estimated by solving of linearized equations for perturbation. In all works the spectrum of perturbation was continuous in longitudinal direction and discrete in the lateral direction – the flow had to be divided into the counting number of paths (braids).

At the stage of linearization of the equations of momentum and continuity for perturbation Callander and Mikhinov used an assumption of the uniform velocity and depth distribution in the main flow. Fredsoe assigned the lateral inclination of the main channel bottom. This approach led to appearance of additional mode of instability at the region of long waves (bars). Solution of Fredsoe shows the importance of the lateral terms of equations for long wave's instability.

This brief analysis of the main works on initial instability in the case of 2D equation of motion and conservation led to following conclusions:

1) All terms in 3D equation of motion and conservation are important for the further instability analysis. The effects of the dynamic pressure vertical distribution and of the helical circulation must be taken into account during the depth-averaging.
2) The form of the equations must allow the perturbation equations linearization with minimal loss of terms.
3) The geometry of initial perturbation have to be assigned in the most general form to obtain the continuous spectrum of waves both in longitudinal and lateral directions.

It was shown (Sidorchuk, 1996) that the most reliable result can be achieved by the analysis of equations of flow momentum, water and sediment mass conservation in curvilinear orthogonal co-ordinates, depth-averaged according to recommendations of Kartvelishvily (1973):

$$\frac{\partial}{\partial t}\int_z^h u dZ + \frac{1}{H_x^2 H_y}\frac{\partial}{\partial X}\left(H_x H_y \int_z^h u^2 dZ\right) + \frac{1}{H_x^2 H_y}\frac{\partial}{\partial Y}\left(H_x^2 \int_z^h uv dZ\right) -$$

$$\frac{1}{H_x^2}\frac{\partial H_x}{\partial X}\int_z^h u^2\mathrm{d}Z-\frac{1}{H_xH_y}\frac{\partial H_y}{\partial X}\int_z^h v^2\mathrm{d}Z+\frac{1}{H_x}\frac{\partial^2}{\partial t\partial X}\int_z^h\int_Z^h w\mathrm{d}\xi\,\mathrm{d}Z$$

$$+\frac{1}{H_x}\frac{\partial}{\partial X}\left\{\frac{1}{H_xH_y}\left[\frac{\partial}{\partial X}\left(H_y\int_z^h\int_Z^h wu\mathrm{d}\xi\,\mathrm{d}Z\right)+\frac{\partial}{\partial Y}\left(H_x\int_z^h\int_Z^h wv\mathrm{d}\xi\,\mathrm{d}Z\right)\right]\right. \quad (1\text{-}1)$$

$$\left.+\frac{1}{H_x}\frac{\partial h}{\partial X}\int_z^h w\,u\mathrm{d}Z+\frac{1}{H_y}\frac{\partial h}{\partial Y}\int_z^h w\,v\mathrm{d}Z\right\}+\frac{1}{H_x}\frac{\partial h}{\partial X}\frac{\partial}{\partial t}\int_z^h w\,\mathrm{d}Z+$$

$$\frac{1}{H_x}\frac{\partial}{\partial X}\int_z^h w^2\mathrm{d}Z+\frac{w^2}{H_x}\frac{\partial h}{\partial X}+\frac{1}{H_x}gd_0\frac{\partial h}{\partial X}+g\frac{\sqrt{(u^2+v^2)}u}{C_0^2}=0$$

$$\frac{\partial}{\partial t}\int_z^h v\mathrm{d}Z+\frac{1}{H_y^2H_x}\frac{\partial}{\partial Y}\left(H_xH_y\int_z^h v^2\mathrm{d}Z\right)+\frac{1}{H_y^2H_x}\frac{\partial}{\partial X}\left(H_y^2\int_z^h uv\mathrm{d}Z\right)-$$

$$\frac{1}{H_y^2}\frac{\partial H_y}{\partial Y}\int_z^h v^2\mathrm{d}Z-\frac{1}{H_xH_y}\frac{\partial H_x}{\partial Y}\int_z^h u^2\mathrm{d}Z+\frac{1}{H_y}\frac{\partial^2}{\partial t\partial Y}\int_z^h\int_Z^h w\mathrm{d}\xi\,\mathrm{d}Z$$

$$+\frac{1}{H_y}\frac{\partial}{\partial Y}\left\{\frac{1}{H_xH_y}\left[\frac{\partial}{\partial Y}\left(H_x\int_z^h\int_Z^h wv\mathrm{d}\xi\,\mathrm{d}Z\right)+\frac{\partial}{\partial X}\left(H_y\int_z^h\int_Z^h wu\mathrm{d}\xi\,\mathrm{d}Z\right)\right]+\right.$$

$$\left.+\frac{1}{H_y}\frac{\partial h}{\partial Y}\int_z^h w\,v\mathrm{d}Z+\frac{1}{H_x}\frac{\partial h}{\partial X}\int_z^h w\,u\mathrm{d}Z\right\}+\frac{1}{H_x}\frac{\partial h}{\partial Y}\frac{\partial}{\partial t}\int_z^h w\,\mathrm{d}Z+$$

$$\frac{1}{H_y}\frac{\partial}{\partial Y}\int_z^h w^2\mathrm{d}Z+\frac{w^2}{H_x}\frac{\partial h}{\partial Y}+\frac{1}{H_y}gd_0\frac{\partial h}{\partial Y}+g\frac{\sqrt{(u^2+v^2)}v}{C_0^2}=0$$

$$\frac{1}{H_xH_y}\left(\frac{\partial}{\partial X}H_y\int_z^h u\mathrm{d}Z+\frac{\partial}{\partial Y}H_x\int_z^h v\mathrm{d}Z\right)+\frac{\partial d_0}{\partial t}=0;$$

$$\frac{1}{H_xH_y}\left(\frac{\partial}{\partial X}H_y\int_z^h\rho\,u\mathrm{d}Z+\frac{\partial}{\partial Z}H_x\int_z^h\rho\,v\mathrm{d}Z\right)+\frac{\partial}{\partial}\frac{(h-d_0)}{t}=0.$$

Natural coordinates s_x and s_y are used; therefore in the further equations the following replacement is valid:

$$s_x = H_x X; s_y = H_y Y; \; K_x = -\frac{1}{H_x}\frac{\partial H_x}{\partial s_y}; \; K_y = -\frac{1}{H_y}\frac{\partial H_y}{\partial s_x}.$$

According to the method of initial instability analysis, the velocities, surface and bottom elevations and channel curvature are presented as a sum of two components: main flow (capital letter) and perturbation (with the asterisk '):

$$u = U + u'; v = v'; w = w'; d_0 = D_0 + d'; h = H + h'; \rho = R; K = K_0 + K' \quad (1\text{-}2).$$

The longitudinal velocity of the main flow is directed along s_x axis, which follows flow-lines, the lateral velocity of the main flow is zero, as well as vertical velocity. The perturbation of the longitudinal velocity also follows flow-lines of the main flow, but lateral velocity perturbation forms a helical circulation along the perturbations of the main flow-lines. In this case the geometry of perturbations of flow-lines must be known.

The Rozovsky-type geometry was used in this study with curvature of the perturbations of longitudinal coordinate related to lateral velocity of helical flow:

$$K'_x = -\gamma_0 \frac{v'}{UD_0}. \quad (1\text{-}3)$$

The coefficient γ_0 is function of Chezy coefficient and flow depth:

$$\gamma_0 = \frac{1 - 7.2\sqrt{D_0}/C_0}{7.4 - 13.0\sqrt{g}/C_0} \quad (1\text{-}4)$$

Usually coefficient γ_0 changes in the range 0.08--0.1.

All terms, which contain depth-averaging integrals, can be replaced by expressions, which include full depth d_0 (or only main flow depth D_0), components (with the star *) of surface or near-bed velocities, near-bed sediment concentration and kinematic coefficients α_i and β_i (which can differ for main flow and perturbation components due to different vertical distribution of these components).

$$\int_z^h u dZ = \left(\alpha_U U^* + \alpha_u u^*\right) d_0; \quad \int_z^h u^2 dZ = \left(\alpha_{U2} U^{*2} + 2\alpha_{uU} u^* U^*\right) d_0;$$

$$\int_z^h v dZ = \left(\alpha_v v^*\right) D_0;$$

$$\int_z^h uv\mathrm{d}Z = \left(\alpha_{Uv} U^* v^*\right) D_0;$$

$$\int_z^h u\rho\mathrm{d}Z = \left(\alpha_{UR} U^* R^* + \alpha_{U\rho} U^* \rho^* + \alpha_{uR} u^* R^*\right) d_0; \tag{1-5}$$

$$\int_z^h v\rho\mathrm{d}Z = \left(\alpha_{vR} v^* R^*\right) D_0;$$

$$\int_z^h uw\mathrm{d}Z = \left(\alpha_U U^* \frac{\partial h'}{\partial t} + \alpha_{U2} U^* \frac{\partial h'}{\partial s_x}\right) D_0;$$

$$\int_z^h \int_Z^h w\mathrm{d}\xi\,\mathrm{d}Z = \left(\frac{\partial h'}{\partial t} + \beta_U U^* \frac{\partial h'}{\partial s_x}\right) D_0^2;$$

$$\int_z^h \int_Z^h uw\mathrm{d}\xi\,\mathrm{d}Z = \left(\beta_U U^* \frac{\partial h'}{\partial t} + \beta_{U2} U^{*2} \frac{\partial h'}{\partial s_x}\right) D_0^2;$$

Here according to linearization procedure the products of perturbations are assumed to be negligibly small.

The kinematic coefficients can be calculated, if vertical distribution of flow components is known. For the main flow the parabolic profile is used for longitudinal velocity

$$U = U^* \left(\frac{Z}{D_0}\right)^n \tag{1-6}$$

and Rouse (1938) formula for the vertical distribution of sediment concentration.

$$R = R_b \left(\frac{D_0 - Z}{Z} \frac{\Delta}{D_0}\right)^{\frac{\omega}{\kappa U_*}} \tag{1-7}$$

The mean sediment concentration in the flow can be calculated after Rossinsky-Kuz'min (1964):

$$R_m = 1.5 * 10^{-5} \frac{U^3}{\omega D_0} \tag{1-8}$$

Much less is known about the flow perturbation. As the perturbation of the lateral velocity forms the helical flow, the linear vertical distribution according Rozovsky (1957) can be used:

$$v' = v^{*}\left(2\frac{Z}{D_0} - 1\right) \tag{1-9}$$

For longitudinal and vertical velocity perturbation the uniform vertical distribution is used.

The perturbation of surface vertical velocity in (1-5) was calculated from the boundary condition on the flow surface:

$$w^{*} = \frac{\partial h'}{\partial t} + u^{*}\frac{\partial h'}{\partial s_x} + v^{*}\frac{\partial h'}{\partial s_y} \tag{1-10}$$

After substitution of (1-2), (1-4) – (1-10) into the system (1-1) and subtraction of the equations for the main flow the following equations of momentum, conservation and deformation for perturbation were obtained:

$$\frac{\partial}{\partial t}\left(\alpha_U U^{*} d' + \alpha_u u^{*} D_0\right) + \frac{\partial}{\partial s_x}\left(\alpha_{U2} U^{*2} d' + 2\alpha_{uU} u^{*} U^{*} D_0\right) - K_y\left(\alpha_{U2} U^{*2} d' + 2\alpha_{uU} u^{*} U^{*} D_0\right)$$

$$- \alpha_{U2} U^{*2} D_0 K'_y + \frac{\partial}{\partial s_y}\left(\alpha_{Uv} U^{*} v^{*} D_0\right) - 2\alpha_{Uv} U^{*} v^{*} D_0 K_x$$

$$+ \frac{\partial^2}{\partial t \partial s_x}\left(\frac{\partial h'}{\partial t} + \beta_U U^{*}\frac{\partial h'}{\partial s_x}\right) D_0^2$$

$$+ \frac{\partial}{\partial s_x}\left[\frac{\partial}{\partial s_x}\left(\beta_U U^{*}\frac{\partial h'}{\partial t} + \beta_{U2} U^{*2}\frac{\partial h'}{\partial s_x}\right) D_0^2 - K_x\left(\beta_U U^{*}\frac{\partial h'}{\partial t} + \beta_{U2} U^{*2}\frac{\partial h'}{\partial s_x}\right) D_0^2\right]$$

$$+ g D_0 \frac{\partial h'}{\partial s_x} + g\frac{\alpha_{U2} U^{*2} d' + 2\alpha_{uU} u^{*} U^{*} D_0}{C_0^2 D_0} = 0$$

$$\frac{\partial}{\partial t}\alpha_v v^{*} D_0 + \frac{\partial}{\partial s_x}\left(\alpha_{Uv} U^{*} v^{*} D_0\right) - 2K_y \alpha_{Uv} U^{*} v^{*} D_0$$

$$+K_x\left(\alpha_{U2}U^{*2}d'+2\alpha_{uU}u^*U^*D_0\right)+\alpha_{U2}U^{*2}D_0K_x'+$$

$$\frac{\partial^2}{\partial t\partial s_y}\left(\frac{\partial h'}{\partial t}+\beta_U U^*\frac{\partial h'}{\partial s_x}\right)D_0^2$$

$$+\frac{\partial}{\partial s_y}\left[\frac{\partial}{\partial s_x}\left(\alpha_U U^*\frac{\partial h'}{\partial t}+\alpha_{U2}U^*\frac{\partial h'}{\partial s_x}\right)D_0-K_y\left(\alpha_U U^*\frac{\partial h'}{\partial t}+\alpha_{U2}U^*\frac{\partial h'}{\partial s_x}\right)D_0\right]$$

$$+gD_0\frac{\partial h'}{\partial s_y}+g\frac{\alpha_{vU}v^*U^*}{C_0^2}=0$$

$$\alpha_u D_0\frac{\partial u^*}{\partial s_x}+\alpha_U d'\frac{\partial U^*}{\partial s_x}+\alpha_u u\frac{\partial D_0}{\partial s_x}+\alpha_U U\frac{\partial d'}{\partial s_x}-\alpha_U U^*D_0K_y'-\alpha_u u^*D_0K_y-\alpha_U d'U^*K_y+$$

$$\alpha_v D_0\frac{\partial v^*}{\partial s_y}+\alpha_v v^*\frac{\partial D_0}{\partial s_y}-\alpha_v v^*D_0K_x+\frac{\partial d'}{\partial t}=0 \qquad (1\text{-}11)$$

$$\alpha_{uR}RD_0\frac{\partial u}{\partial s_x}+\alpha_{UR}Rd'\frac{\partial U}{\partial s_x}+\alpha_{uR}Ru\frac{\partial D_0}{\partial s_x}+\alpha_{UR}RU\frac{\partial d'}{\partial s_x}+\alpha_{UR}d'U\frac{\partial R}{\partial s_x}+\alpha_{uR}D_0u\frac{\partial R}{\partial s_x}$$

$$-\alpha_{UR}RUD_0K_y'-\alpha_{uR}RuD_0K_y-\alpha_{UR}Rd'UK_y$$

$$+\alpha_{vR}RD_0\frac{\partial v}{\partial s_y}+\alpha_{vR}Rv\frac{\partial D_0}{\partial s_y}+\alpha_{vR}D_0v\frac{\partial R}{\partial s_y}-\alpha_{vR}RvD_0K_x+(1-\varepsilon)\frac{\partial\left(h'-d'\right)}{\partial t}=0$$

The products and powers of the perturbation components were excluded from the system (1-11) due to linearization. The use of curvilinear co-ordinates allows keeping in the linearized equations the terms with curvature of the longitudinal and lateral coordinates, to which lateral movement is related. The curvature of the lateral coordinate perturbation is assumed to be a function of the curvature of the longitudinal coordinate perturbation:

$$K_y'\approx\frac{k_x}{k_y}K_x' \qquad (1\text{-}12)$$

The initial perturbations of the flow components, surface and channel bed elevations were presented in the form of longitudinal moving sinusoidal wave with complex celerity

$$c = c_R + ic_i:$$

$$u^* = A(s_y)\exp[ik_x(s_x - ct)]; \quad v^* = B(s_y)\exp[ik_x(s_x - ct)];$$

$$h^* = P(s_y)\exp[ik_x(s_x - ct)]; \quad (h^* - d^*) = T(s_y)\exp[ik_x(s_x - ct)]. \tag{1-13}$$

The functions *A*, *B*, *P* and *T*, which describe the perturbation shape in lateral direction, are also complex. It means that the perturbation can be composed of two moving sinusoidal waves with different initial amplitude and phase (only the real part of (1-13) is taken into consideration). In this case the geometry of the longitudinal coordinate perturbation is described by sine-generated curve, which is used for meandering channel shape description (Langbein, Leopold, 1966).

After substitution of (1-13), (1-3) and (1-12) into (1-11), the system of equations of momentum and conservation is rewritten in form

$$a_1A + b_{10}\frac{\partial B}{\partial s_y} + b_1B + c_1P + d_1T = 0;$$

$$a_2A + b_2B \quad + c_{20}\frac{\partial P}{\partial s_y} + c_2P + d_2T = 0; \tag{1-14}$$

$$a_3A + b_{30}\frac{\partial B}{\partial s_y} + b_3B + c_3P + d_3T = 0$$

$$a_4A + b_{40}\frac{\partial B}{\partial s_y} + b_4B + c_4P + d_4T = 0.$$

Here:

$$a_1 = 2\alpha_{uU}Uik_x - \alpha_u ik_x c + 2g\alpha_{uU}\frac{U}{C_0^2D_0} + \alpha_u\frac{1}{D_0}\frac{\partial}{\partial t}D_0 + 2\alpha_{uU}\frac{\partial}{\partial s_x}U + 2\alpha_{uU}\frac{U}{D_0}\frac{\partial}{\partial s_x}D_0$$

$$- 2\alpha_{uU}UK_y$$

$$b_{10} = \alpha_{vU}U$$

$$b_1 = \frac{1}{D_0}\frac{\partial}{\partial s_y}(\alpha_{Uv}UD_0) - 2\alpha_{Uv}UK_x;$$

$$c_1 = gik_x - \alpha_U \frac{U}{D_0} ik_x c + g\alpha_{U2} \frac{U^2}{C_0^2 D_0} + \alpha_{U2} \frac{U^2}{D_0} ik_x - K_y \alpha_{U2} \frac{U^2}{D_0}$$

$$- D_0\left(ik_x^3 c^2 - 2U\beta_U ik_x^3 c + \beta_{U2} ik_x^3 + K_x \beta_U k_x^2 c - K_x \beta_{U2} k_x^2\right) - \frac{\left(ik_x c - \beta_U ik_x U\right)}{D_0} \frac{\partial^2}{\partial t \partial s_x} D_0^2 \tag{1-15}$$

$$- \frac{\beta_U Uik_x c + \beta_{U2} U^2 ik_x}{D_0} \frac{\partial^2}{\partial s_x^2} D_0^2 + \frac{\beta_U Uik_x c + \beta_{U2} U^2 ik_x}{D_0} \frac{\partial}{\partial s_x} K_x D_0^2$$

$$d_1 = \alpha_U \frac{U}{D_0} ik_x c - g\alpha_{U2} \frac{U^2}{C_0^2 D_0} - \alpha_{U2} \frac{U^2}{D_0} ik_x + K_y \alpha_{U2} \frac{U^2}{D_0}$$

$$a_2 = K_x 2\alpha_{U2} U$$

$$b_2 = \alpha_{Uv} Uik_x - \alpha_v ik_x c + g\alpha_{vU} \frac{U}{C_0^2 D_0} - \alpha_{U2}\gamma_0 \frac{U}{D_0} + \alpha_v \frac{1}{D_0} \frac{\partial D_0}{\partial t} + \alpha_{Uv} \frac{U}{D_0} \frac{\partial D_0}{\partial s_x} + \alpha_{Uv} \frac{\partial U}{\partial s_x} - 2K_y \alpha_{Uv} U$$

$$c_{20} = g - D_0\left(k_x^2 c^2 - 2U\beta_U k_x^2 c + \beta_{U2} k_x^2 - K_x \beta_U ik_x c + K_x \beta_{U2} ik_x\right)$$

$$c_2 = -K_x \alpha_{U2} \frac{U^2}{D_0} - \frac{\left(ik_x c - \beta_U ik_x U\right)}{D_0} \frac{\partial^2}{\partial t \partial s_y} D_0^2$$

$$- \frac{\beta_U Uik_x c + \beta_{U2} U^2 ik_x}{D_0} \frac{\partial^2}{\partial s_x \partial s_y} D_0^2 + \frac{\beta_U Uik_x c + \beta_{U2} U^2 ik_x}{D_0} \frac{\partial}{\partial s_y} K_y D_0^2$$

$$d_2 = K_x \alpha_{U2} \frac{U^2}{D_0}$$

$$a_3 = \alpha_u D_0 ik_x + \alpha_u \frac{\partial D_0}{\partial s_x} - \alpha_u D_0 K_y;$$

$$b_{30} = \alpha_v D_0;$$

$$b_3 = \alpha_v \frac{\partial D_0}{\partial s_y} - \alpha_v D_0 K_x$$

$$c_3 = \alpha_U Uik_x - ik_x c + \alpha_U \frac{\partial U}{\partial s_x} - \alpha_U UK_y;$$

$$d_3 = -\alpha_0 Uik_x + ik_x c - \alpha_0 \frac{\partial U}{\partial s_x} + \alpha_U UK_y;$$

$$a_4 = \alpha_{uR} RD_0 ik_x + \alpha_{uR} R \frac{\partial D_0}{\partial s_x} + \alpha_{uR} D_0 \frac{\partial R}{\partial s_x} - \alpha_{uR} RD_0 K_y$$

$$b_4 = \alpha_{vR} R \frac{\partial D_0}{\partial s_y} + \alpha_{vR} D_0 \frac{\partial R}{\partial s_y} - \alpha_{vR} RD_0 K_x$$

$$b_{40} = \alpha_{vR} RD_0$$

$$c_4 = \alpha_{UR} R \frac{\partial U}{\partial s_x} + \alpha_{UR} RUik_x + \alpha_{UR} U \frac{\partial R}{\partial s_x} - \alpha_{UR} RUK_y$$

$$d_4 = -ik_x c(1-\varepsilon) - \alpha_{UR} R \frac{\partial U}{\partial s_x} - \alpha_{UR} RUik_x - \alpha_{UR} U \frac{\partial R}{\partial s_x} + \alpha_{UR} RUK_y$$

Note that all these coefficients do not change along s_y.

When A, B and T are eliminated from (1-14) the result is an ordinary second order differential equation (the equation of damped oscillations) for the lateral perturbations of flow surface:

$$\frac{d^2 P}{ds_y^2} + 2\delta \frac{dP}{ds_y} + k_y^2 P = 0; \qquad (1\text{-}16).$$

or

$$\frac{d^2 \theta}{ds_y^2} + \left(k_y^2 - \delta^2\right)\theta = 0; \qquad (1\text{-}17)$$

Here

$$\theta = P \exp\left(-\delta s_y\right)$$

The boundary conditions on the channel banks for this function of flow surface elevations can be non-zero. This means that the eigenvalues of equation (1-16) are continuous

$$k_0^2 = \left(\frac{\nu\pi}{W}\right)^2 = k_y^2 - \delta^2 \qquad (1\text{-}18)$$

Here ν runs through all positive real values. The use of non-zero boundary conditions leads to continuous spectrum of perturbation waves both in lateral and longitudinal direction.

The eigenfunctions of (1-16) will have the form of

$$P \sim \exp(-\delta s_y)\cos k_0\left(s_y - \frac{W}{2}\right) \quad (1\text{-}19).$$

The system (1-14) was analysed by Mikhinov (1983). He showed that there are three types of waves in a flow and on the bottom of a channel: 1) waves with rate of motion $c_r >> U$; 2) two waves (direct and indirect) with $c_r \sim U$; 3) slow waves with rate of motion $c_r << U$. The first two types of waves cannot be the cause of channel relief generation due to their high rate of motion. Only the third type of perturbation waves with low rates of motion along the channel can generate the channel relief. This class of waves does not appear in solution of system (1-14) written without the equation of deformation.

The system was simplified with the use of condition $\partial / \partial t = 0$ (in the momentum equations in (1-11)) or c=0 (in the coefficients with the indices 1 and 2 in (1-14 and 1-15)). In this case waves of the first and second types are both eliminated, and only slow waves of the third type are analysed.

With these simplifications the solution of (1-16) takes the form of the dispersion equation

$$A_1 c^2 + A_2 c + A_3 = 0 \quad (1\text{-}20)$$

The coefficients A_i are the functions of parameters of the main flow, kinematic coefficients, longitudinal and lateral wave-numbers.

The solution of system (1-14) depends on the geometry of the main flow. Three flow configurations are analysed: 1) main flow is straight with the uniform velocity and depth; 2) main flow is straight with the non-uniform longitudinal velocity and depth in longitudinal direction, but uniform in lateral direction; 3) main flow is bended, with the non-uniform longitudinal velocity and depth in both direction.

Straight Main Flow with the Uniform Longitudinal Velocity and Depth

In this case all derivatives of U, R and D_0, as well as main flow-line curvatures are zeros and (1-14) take the form

$$\begin{aligned}
&a_1 A + b_{10}\frac{\partial B}{\partial s_y} + c_1 P + d_1 T = 0;\\
&b_2 B + c_{20}\frac{\partial P}{\partial s_y} = 0;\\
&a_3 A + b_{30}\frac{\partial B}{s_y} + c_3 P + d_3 T = 0;\\
&a_4 A + b_{40}\frac{\partial B}{s_y} + c_4 P + d_4 T = 0.
\end{aligned} \quad (1\text{-}21)$$

In this case δ=0 and (1-17) takes the form of the equation governing a simple pendulum

$$\frac{d^2P}{ds_y^2}+k_y^2P=0 \tag{1-22}$$

with continuous eigenvalues:

$$k_y^2=\left(\frac{\nu\pi}{W}\right)^2 \tag{1-23}$$

and the eigenfunctions

$$P\sim \cos k_y\left(s_y-\frac{W}{2}\right) \tag{1-24}$$

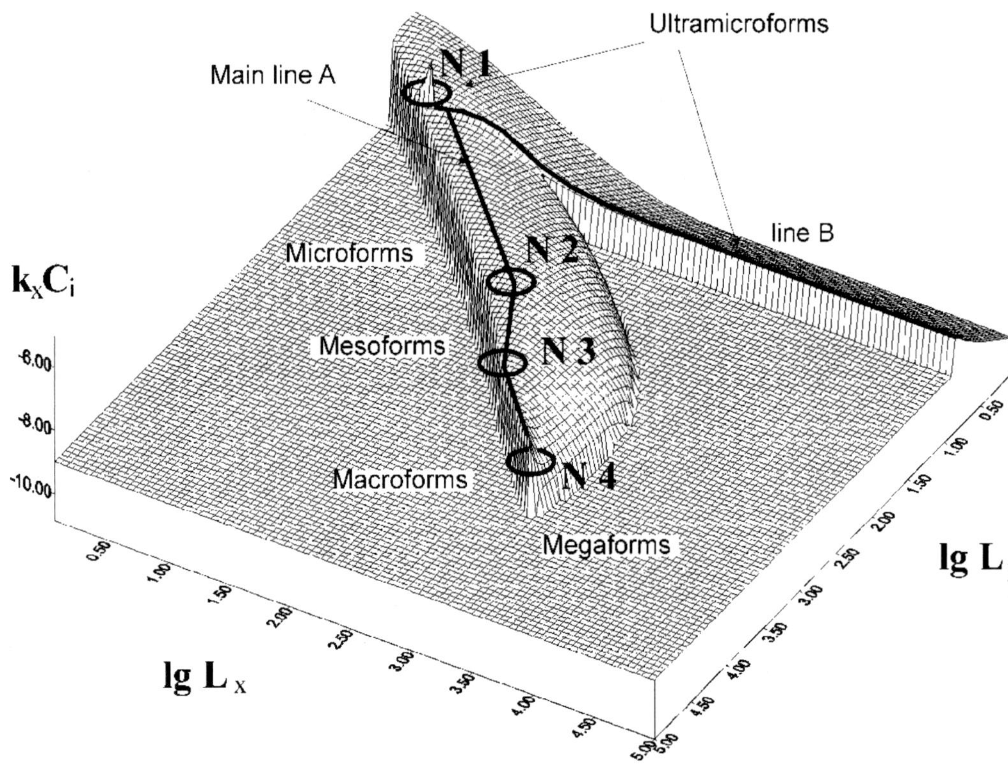

Figure 1.1. Two-dimensional theoretical spectrum of unstable waves amplitude increase for the case of U=2.0 m/s; D_0=5 m; C_0=40. N1, N2, N3, N4 are boundary points between fields of ultramicroforms, microforms, mesoforms, macroforms, and megaforms.

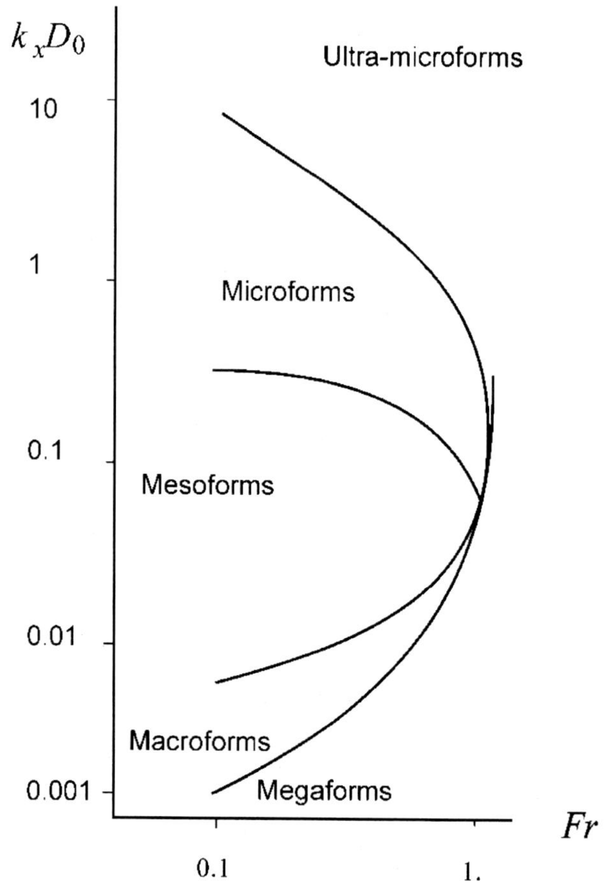

Figure 1.2. The main fields of the theoretical types of river bed undulations.

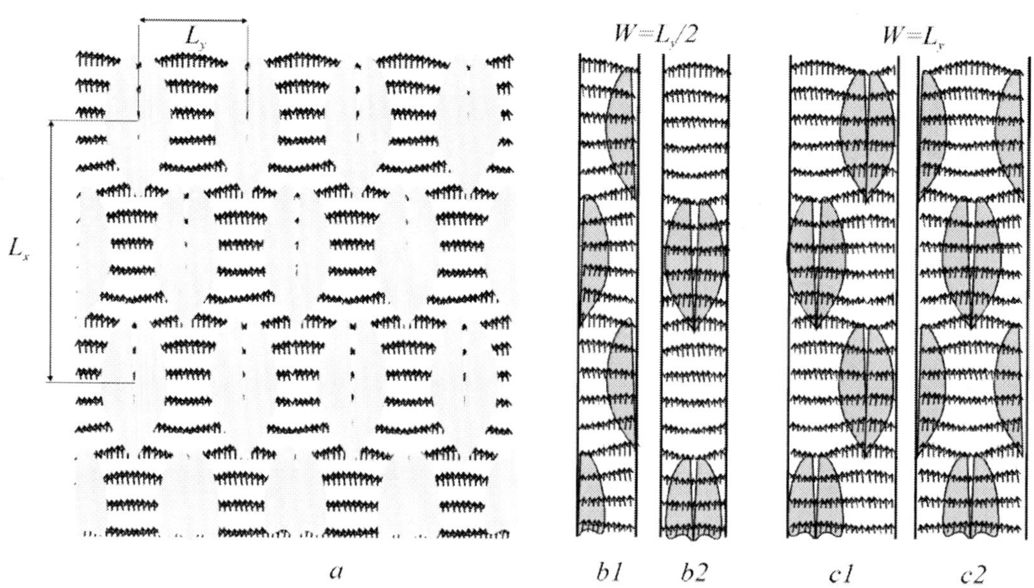

Figure 1.3. Geometry of the river bed theoretical undulations.

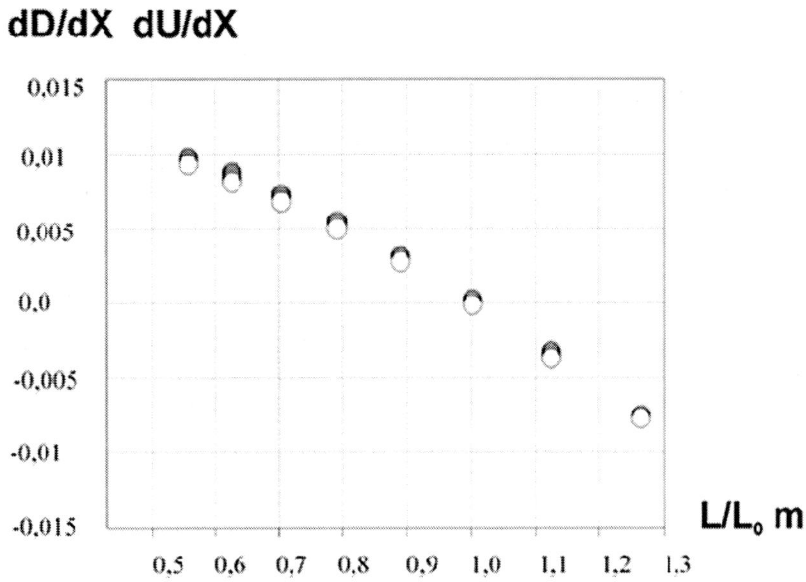

Figure 1.4. Relationship between the theoretical mesoform length and the main flow velocity and depth longitudinal gradients.

The dispersion equation, which follows from (1-22), was solved for the wide range of flow parameters and wave-number values. The solutions can be represented in the form of 2D spectra of the decrements of exponential growth of the wave's amplitude c_i (Figure 1.1). This spectrum is continuous both in longitudinal and lateral directions due to non-zero boundary conditions. Theoretical spectrum of unstable channel bed waves consists of a broad range of wave-numbers. Therefore, alluvial relief with such spectrum is highly hierarchical: smaller waves complicate the larger ones, which transform even larger waves, and so on. At the same time, the spectrum surface is non-even, several regions and fields of specific relations between c_i , k_x and k_y can be defined.

Two main regions are defined by c_i sign. In the region of positive c_i the amplitudes of initial perturbations increase in time; these undulations are unstable and potentially can form the fluvial relief. In the region of negative c_i the amplitudes of initial perturbations decrease in time; these undulations are stable and cannot form the fluvial relief.

Inside the region of unstable waves, several special fields, lines and points were defined. The main line is the line with local maximums of $k_1 c_i$ (Figure 1.1, line A). Beginning from Kennedy's (1963) study of the channel bed initial instability, the unstable waves with the maximum growth rate of amplitude $k_x c_i$ are assumed to be responsible for bed relief formation. There are four special points on the main line: N1 - the local maximum of amplification of short waves; N2 - the point of the beginning of rapid increase in elongation of the wave's shape; N3 - the local minimum of amplification of the long waves; N4 - the local maximum of the long waves' amplification. The imagined boundaries, which pass through these points, divide the region of unstable waves into five fields:

1) The field of very short waves (ultra-microforms). Two main types of such waves are defined: 2D and 3D ultra-microforms. Line B on Figure 1.1 shows the maximum of

2D ultra-microforms amplification in the spectrum. Mikhinov (1983) derived the formula for calculation of the lengths of these waves

$$L_x = 5.4DFr \qquad (1\text{-}25).$$

The length of 2D ultra-microforms in lateral direction is infinite.

Point N1 on the main line represents the most probable location of 3D ultra-microforms in the spectrum. Their lengths are about 20% larger than those of 2D ultra-microforms, but their type of relation with the depth and Froude number is the same for $Fr<0.5$ (Figure 1.2). The shape of most probable 3D ultra-microforms on the horizontal plane is isometric.

2) The field of short waves (microforms).The shape of the most probable 3D microforms (on the main line of spectrum) is symmetrical on the horizontal plane. There are no local maximums in the spectrum within the field of microforms, the rate of wave's amplitude growth decreasing steadily with longitudinal wave-number.
3) The field of middle-size waves - mesoforms.The most probable mesoforms on the main line of spectrum are also 3D, but their shape on the horizontal plane is elongated. The position of the boundary point N2 between the fields of symmetrical microforms and elongated mesoforms (i.e. the longitudinal wavelength of the most probable boundary undulation) changes with depth, Froude number and resistance to flow (Figure 1.2):

$$L_x = 5.2D_0\left(\frac{C_0^2}{2g}\right)^{0.3}\exp(2.0Fr) \qquad (1\text{-}26).$$

The rate of wave's amplitude growth at this field also decreases steadily with longitudinal wavelength increase and reaches it minimum at the boundary point N3. The position of boundary point N3 changes with depth, Froude number and resistance to flow (Figure 1.2):

$$L_x = 8.0D_0\left(\frac{C_0^2}{2g}\right)\exp(-1.8Fr) \qquad (1\text{-}27).$$

Note that within the field of mesoforms the relationship between the longitudinal wavelength of the most probable undulations and Froude number changes from direct to inversed.

4) The field of long waves (macroforms). The most probable macroforms on the main line of spectrum are 3D, with elongated shape on the horizontal plane. The rate of wave's amplitude growth at this field increases steadily with longitudinal wavelength and reaches its maximum at the point N4 on the main line of spectrum. The lengths of the long waves in this point increase with main channel depth D_0 and Chezy coefficient value, but decrease with Froude number increase (Figure 1.2):

$$L_x = 4.2D_0\left(\frac{C_0^2}{2g}\right) * Fr^{-1.1} \qquad 0.1 \le Fr \le 0.5$$

$$L_x = 42.0D_0\left(\frac{C_0^2}{2g}\right)\exp(-3.1Fr) \qquad 0.5 < Fr < 1.0 \tag{1-28}$$

The degree of elongation of the macroforms at this point is related to intensity of helical circulation and to channel resistance:

$$L_x / L_y = \sqrt{\frac{C_0^2 \gamma_0}{g}} \tag{1-29}$$

5) The field of very long waves (megaforms). On the main line of the spectrum the width of these forms is very small in relation to their length. The rate of amplitude growth is relatively low in the field of these waves. Typically, megaforms have a shape of long and narrow strips in the channel.

Initial undulation of the channel bed has the same bi-sinusoidal geometry at the horizontal plane, as undulations of the flow surface (Figure 1.3a):

$$T \sim \sin(k_x s_x)\cos k_y\left(s_y - \frac{W}{2}\right) \tag{1-30}$$

There are no restrictions for the boundary conditions for (1-30); therefore river channel banks (traced along coordinate s_x at the lateral coordinates b_0 and b_0+W) can isolate an endless variety of channel geometry types, each controlled by the combination of channel width W, longitudinal L_x and lateral L_y wavelength and initial lateral phase $\frac{2\pi b_0}{L_y}$. For example, for the macroforms these combinations produce all the main types of the real river channels. If the channel width is about one half of the lateral wave-length (Figure 1.3b), the channel is "meandering" or "straight with alternating bars" (asymmetric configuration *b1*) or "straight with linguoid or braid bars" (symmetric configuration *b2*). If the channel width is close to the lateral wavelength (Figure 1.3c), the channel is "with alternating islands" (asymmetric configuration *c1*) or "braided" (symmetric configuration *c2*). This simplified picture can be complicated with the smaller undulations from the continuous-discrete spectrum to obtain the hierarchical system of alluvial features – bed forms and bars.

Straight Main Flow with Non-Uniform Longitudinal Velocity and Depth

In this case derivatives $\partial/\partial s_y$, as well as main flow-line curvatures, are zeros in (1-15). The system (1-14) takes the same form (1-21), as in previous case, with changed coefficients a_1, c_1, b_2, a_3, c_3, d_3, a_4, c_4, d_4 due to existence of $\partial/\partial s_x$ derivatives.

These derivatives change the position of the boundaries between the main fields of unstable undulations. The boundary point N2 between fields of microforms and mesoforms is shifted towards the longer wavelengths if the main flow is slowing downstream and vice versa. The same result follows when the flow depth decreases downstream. The field of microforms, on the contrary, becomes narrower in accelerating and deepening flow (Figure 1.4).

Bent Main Flow with the Non-uniform Velocity and Depth

In this case full system (1-14) and (1-15) must be analysed. The main effect of lateral variability of the main flow is the accompanying lateral asymmetry of perturbations.

PART 2

Morphology of Alluvial Features in the Natural Flows: Determination of Alluvial Features

There is an opinion (Jain and Kennedy, 1974, Coleman and Melville, 1996) that the initial instability of the channel bed, which is described by the initial instability analysis, results in some kind of initial-stage bed forms – sand-wavelets, which cannot be identified with mature bed forms in the real river channels. Special experiments were performed to find these initial-stage bed forms. This point of view is true with respect to the amplitudes of the alluvial features, as the initial instability analysis of linearized equations cannot predict the finite amplitude of undulation in principle. As to the size of the undulation at the horizontal plane – their length and width – the question is more difficult. According to the initial instability analysis, the length of initial undulation of the channel bed is equal to the length of the initial wave in the flow. This pair is already in an equilibrium, which is characteristic for active and mature bed form, but not for transforming sand-wavelets. One of the methods to validate the results of the initial instability analysis is to compare them with the results of the measurements in the real river channels in the broad range of geology, relief and climatic conditions.

A natural river channel alluvial relief is hierarchical (Figure 2.1), with a broad range of alluvial forms and bed forms sizes. Step-by-step interpolation technique was used for detection of different forms on the channel bed long profile (Sidorchuk, 1996). The algorithm is as follows (Figure 2.1): 1) A continuous long profile of channel bed altitudes z with all the hierarchy of alluvial forms is digitised with step dx, which is 0.1-0.2 of the smallest detectable bed form length. If a digital sounder was used, its time resolution should meet these conditions. 2) The gradients of altitudes dz/dx of this line of the first order are calculated. 3) Sink points (where the sign of this gradient changes from minus (zero) to plus) are connected by the line of second order. 4) The geometry (length, height, area, asymmetry, etc.) of positive bed forms above this line is measured and listed as for alluvial features of the first order. 5) Bed forms of the first order are 'removed' from the profile, so that the line of the second order becomes a 'surface' line. Steps 2-5 are repeated until all wavelike features are measured and removed from the profile and the shape of the channel main bed (negative form) is estimated.

Step-by-step interpolation technique for bed forms differ the zero-crossing method (Carling and Orr, 2000) as only one negative form is defined in the end – the main channel bed without the positive alluvial features (sand waves). If the main channel bed is not covered completely by mobile alluvium and some parts of it are exposed, these parts should be distinguished before step-by-step interpolation. Usually, some textural or lithological contacts occur at the surface of the main channel bed, and such contacts can be estimated via geophysical methods or direct coring. Alluvial features, separated by stretches of exposed channel bed are called "deficit" forms, and their morphology and dynamics are different from those of continuous alluvial features.

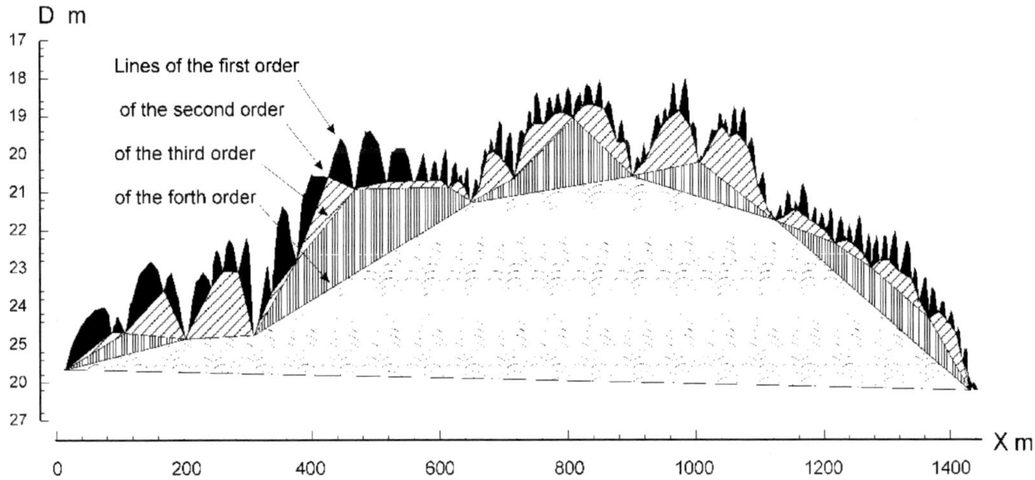

Figure 2.1. Subdivision of alluvial features at the lower Volga River channel with the use of step-by-step interpolation technique. Dunes of the orders 1, 2, and 3 are distinguished on the first order bar.

2D long sections of 3D features can contain quasi-hierarchy due to various deviations of the axes of alluvial features from the long-section direction. To avoid errors related to this phenomenon, the alluvial features hierarchy is investigated in dynamics, using repeated measurements of the same long section at different times. Quasi-dunes disappear from such sequential profiles and hierarchy of alluvial features can be estimated more accurately. Nevertheless, step-by-step interpolation always produces some level of noise, so the quantity of measured alluvial features must be large enough for statistically correct results.

The results of step-by-step interpolation are usually displayed as probability distribution functions (PDFs) of morphological characteristics of alluvial features of different hierarchic levels. The 'tails' of these PDFs often contain some noise, but the modal and mean values are usually statistically significant, and the error of estimate is less than 5% for a number of observations more than 100.

Zero-crossing method with step-by-step interpolation can be used for ordering the river channel meanders (Sinnok and Rao, 1984) and braids.

Another method of alluvial relief ordering is spectral analysis of the altitudes on the channel bed long profiles. To avoid too long series it is practical to use a set of discrete realisations of the bed altitudes, measured on the same profile with different spatial steps *dx*. Each wave system with the wavelength L can be characterised by profiles 100-200L long, digitised with dx=0.1-0.2L. Spatial series of different resolution are transformed with one of the spectral analysis techniques (usually by Fast Fourier Transform) into a set of amplitude-wavenumber spectra. Each individual spectrum contains information about alluvial forms with the length more than $2dx$. Then the individual spectra are linked to form a complex one, which shows distribution of bottom waves on the whole range of lengths. Usually these spectra are characterised by several modes (locations) of amplitudes; each of them can be correlated with alluvial feature of a definite order (Figure 2.2).

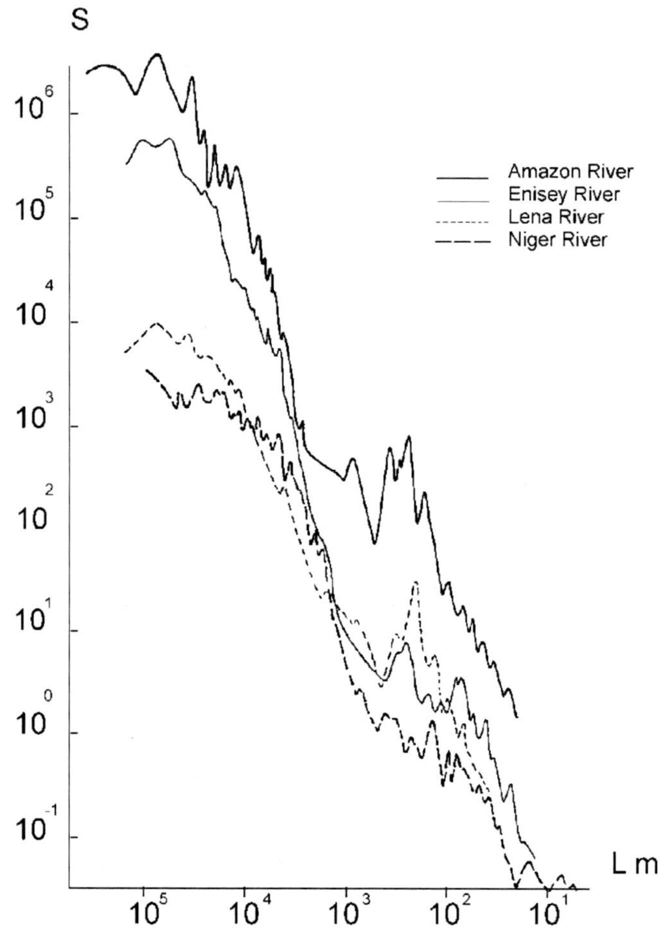

Figure 2.2. Power spectrums of bottom altitudes at lower reaches of several large rivers.

Trigonometric functions are used in a common spectral analysis. The shapes of alluvial features are rarely described properly with these functions. Therefore the modes (locations) on spectra are rather broadband, and often false (multiple) modes are generated.

The method of step-by-step interpolation usually gives more information about morphometric parameters of individual alluvial forms, about their mean values and statistical distributions. Spectral analysis provides a direct estimate of general characteristics of the association of alluvial forms, and presumably such estimates are closer to the ones obtained using a theoretical approach. Both methods produce significant noise, but their joint-use gives a possibility of cross-control and help to avoid large errors (Figure 2.3).

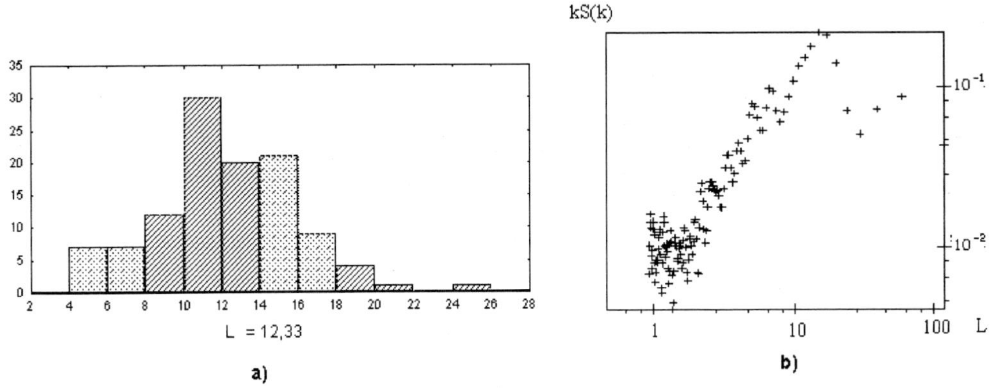

Figure 2.3. Comparison of the results of morphometric analysis of the Vychegda River bed forms with histogram and power spectrum methods.

The hierarchical structure of river channel relief is obvious also on detailed river maps, air-photos and sonar images of river channels. Generally these sources contain much more information about three-dimensional geometry of alluvial forms than bottom profiles. But complexity of hierarchical structure is seldom fully visible on them. River topographical maps usually do not show small bed forms. On the air-photos steep small bottom forms can be detected at the shallows, but it is difficult to correlate their size with hydraulic parameters of the time of their origination. Side-looking sonar images, coupled with the digital bottom sounding, are the most promising sources of information about 3-D bottom relief geometry.

Morphology of the Hierarchical Systems of Alluvial Features at Different Rivers

Investigations of alluvial relief hierarchy were performed during the field works at the lower reaches of the largest rivers of the Western and Eastern Siberia: Yenisei, Ob' with tributaries, Lena, Yana; on the large and medium rivers of the East European plain: Volga, Vychegda with tributaries, Oka, Terek, Dniestr; as well as at the Niger and Benue rivers in the Western Africa (Table 1). Some information was collected from publications and from the results of experiments in the large flumes. As these data were mainly collected at the large rivers with sandy alluvium, the results and conclusions relate to this class of rivers.

As it was already mentioned, the most general classification of alluvial features was proposed by ASCE Task Committee (Nomenclature..., 1966): ripples, dunes and bars. Richards (1980) proposed a term "megaripples" for the bed forms smaller than dunes, but with the size changing with channel depth. In a few publications each of these broad classes of alluvial features was subdivided into several orders of forms (see, for example, Allen, 1978). Observations at the large and middle-size rivers with sandy alluvium show a complicated hierarchy of alluvial features (table 2-1), which usually includes from six to nine levels: megaripples, dunes of 2 or 3 orders, bars of 2 or 3 orders, meanders or/and islands, and complexes of meanders and islands (Sidorchuk, 1996). These features are characterised by distinct morphology and dynamics, which are similar within the same hierarchic level and are different at different hierarchic levels.

Table 1. Hierarchy of the channel relief for the rivers with sandy alluvium

N	River	Station	Discharge m^3/s		U m/s	D m		Type of alluvial feature							
			mean	mean maximum	for a flood period			meander	bar_3	bar_2	bar_1	dune_3	dune_2	dune_1	ripple
1	Enisey	Igarka	17800	110000	1.2	30.	L	25000		10000	5600	2400	400	180	50
							C_v	0.26		0.21			0.31		
2	Lena	Yakutsk					L		45000	14500	4500	800	450	140	5
2	Volga	Astrakhan'	7240	28000	1.2	17.0	L		13000	5900	3100	580	140	43	5
							C_v			0.24	0.28	0.53	0.59	0.72	
3	Northern Dvina	Abramtsevo	2000	12000			L		15000	8000	3500	1330	300	70	
							C_v			0.16	0.30	0.52			
4	Niger	Lokodja	6100	21100	1.5	15.0	L	16100	5100	4900	1300	700	270	40	6
							C_v	0.36	0.23	0.22	0.27	0.32	0.30	0.41	0.40
5	Ob'	Fominskoye	1170	5440	1.2	3.4	L	10100	5900	3000	1100		250	30	3
							C_v	0.22		0.21	0.34		0.43		0.42
6	Pechora	Mitrofanovo	510	5050	1.5	6.0	L	7500		1800	400		60	12	
							C_v								
7	Oka	Puschino	360	5300	1.2	8.0	L	2400		1750	290	135	65	30	
							C_v			0.31	0.27	0.33	0.2	0.27	
8	Vaga	Shegovary	340	3160	1.3	6.0	L	2100			760	210	40	10	
							C_v								
9	Dniestr	Bendery	320				L			1800	390		70	12	
							C_v								
10	Terek	Stepnoye	300	1020	2.5	4.0	L	2400		1230	280		90	30	6
							C_v	0.16		0.26	0.3		0.3	0.4	0.45
11	Vashka	Vedinga	65	640	1.2	4.7	L	1500		1200	230	100	30	10	

N	River	Station	Discharge m^3/s		U m/s	D m		Type of alluvial feature							
			mean	mean maximum	for a flood period			meander	bar_3	bar_2	bar_1	dune_3	dune_2	dune_1	ripple
12	Vychegda	834-847 km	69				L	510	345	90		15			
13	Vychegda	817-834 km	80				L	540	440	100		15			
14	Vychegda	767-780 km	141				L	1000	385	140		40			
15	Vychegda	754-763 km	141				L	1600	650	215		45			
16	Vychegda	544-520 km	355				L	3850	1500	320		120			
17	Vychegda	469-488 km	427				L	3650	1600	500		115			
18	Vychegda	377-400 km	600			4.7	L	7300	2100	950		275			
19	Vychegda	298-331 km	630			4.9	L	6200	2900	700		180			
20	Vychegda	247-269 km	891			7.3	L	3200	550	270		75			
21	Vychegda	197-218 km	933			6.3	L	3400	460	500		125			
22	Vychegda	40-70 km	1122			5.5	L	4400	1580	500					
23	Sysola	269-240 km	50				L	460	290	90		20			
							C_v	0.24	0.23	0.3		0.42			
24	Sysola	220-192 km					L	760	260	90		30			
							C_v	0.20	0.18	0.30		0.18			
25	Sysola	40-70 km					L	1150	315	70		25			
							C_v	0.12	0.22	0.28		0.23			
26	Vishera	12-50 km	80				L	950	380	64					
							C_v	0.21	0.17	0.32					
27	Nivshera	14-54 km	40				L	420	320	55					
							C_v	0.19	0.23	0.41					

* Meander and bar length L was measured between adjacent crosses, dunes and ripples length - between adjacent hollows. C_v is variation coefficient σ/L.

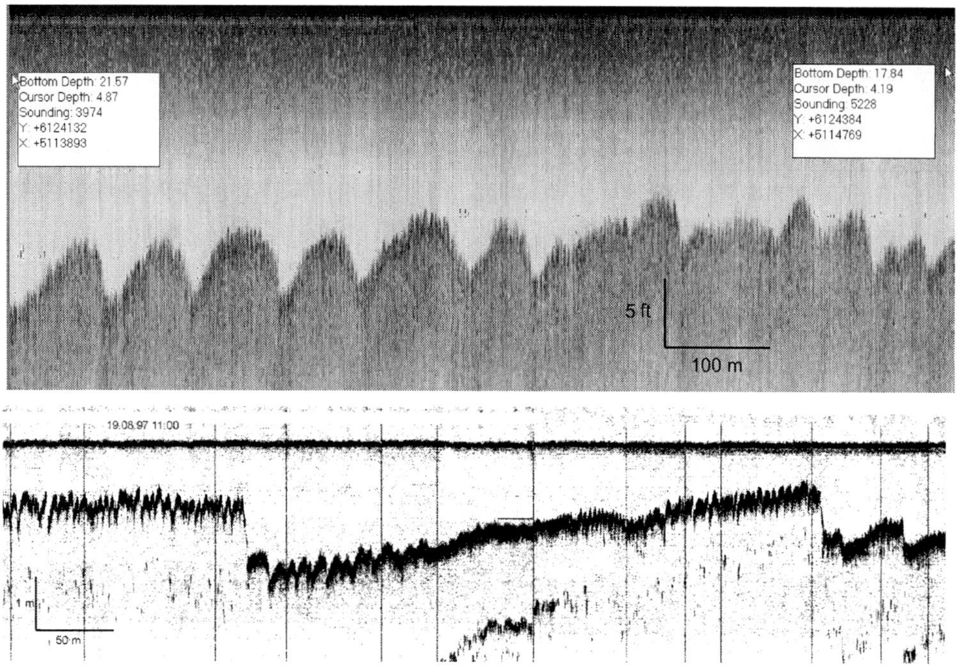

Figure 2.4. Megaripples on dunes at the lower Volga River during the flood (top) and at the lower Vychegda River during the low water period (bottom). Note: the vertical scale at the top figure is in feet.

Megaripples

Megaripples are two- and three-dimensional bed forms, recognized on precise bathymetric profiles and side-looking sonar images. Morphology and dynamics of megaripples are mainly known from numerous experiments in flumes. In the descriptions of such experiments the term "ripples" is traditionally used, however, the size of these features usually changes with flow characteristics. The size of the true ripple, by definition (Nomenclature..., 1966), is independent of flow depth. Here we mostly describe observations at several large rivers (lower Niger, middle Ob', middle Lena, lower Terek, and lower Vychegda), where megaripples are clear enough to be defined on the sonar profiles (Figure 2.4, Table 2).

Wang and Chen (1980) showed that distribution of ripple lengths in a flume fits to gamma-distribution. The same two-parameter gamma-distribution can be used for megaripples in the rivers (Figure 2.5a):

$$dp = \frac{\beta^{\alpha}}{\Gamma(\alpha)} L^{\alpha-1} \exp(-\beta L) dL \qquad (2\text{-}1)$$

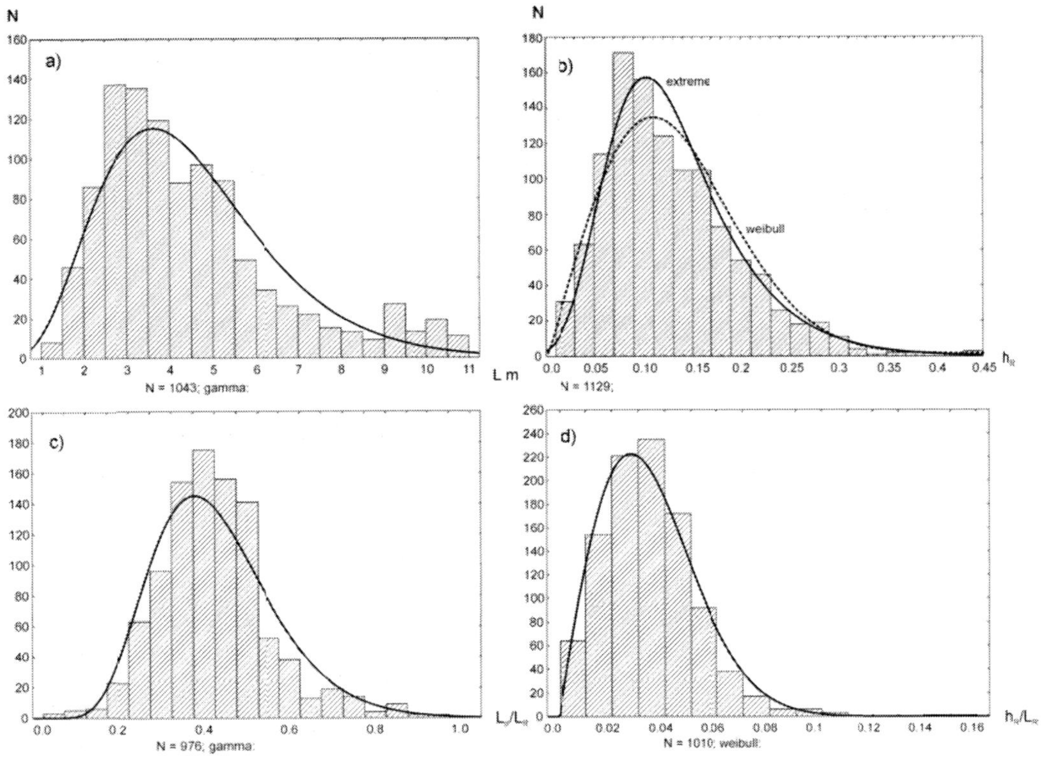

Figure 2.5. Histograms of megaripple geometrical variables with fitting functions (a – length; b – height; c – asymmetry; d – steepness) at the lower Vychegda River during the low water period. Here N is the number of observations.

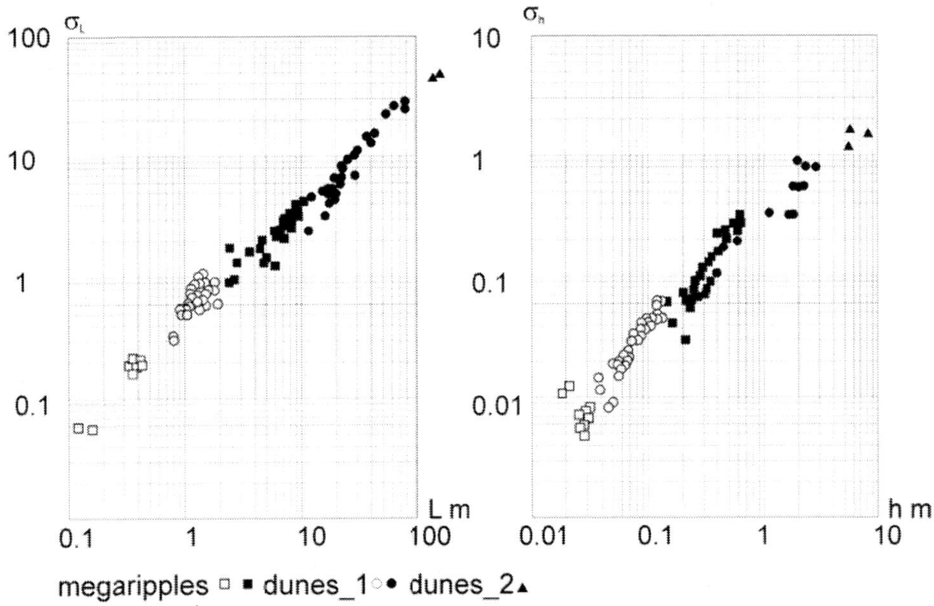

Figure 2.6. Increase in standard deviation of ripple and dune lengths and heights with the increase of their mean sizes. Open symbols refer to the experiments in large flumes, filled symbols refer to the data from large rivers with sand alluvium.

Table 2. Length L and height h and their standard deviations σ for megaripples and dunes_1 in the river channels (*U* and *D* are flow velocity and depth at the middle of the measurements interval, *n* is the number of measured bedforms)

U m/s	*D m*	L_d	σ	h_d	σ	*n*	L_R	σ	h_R	σ	*n*
River Niger, section Lokodja-Adjakuta, flood 1978, slope S=0.1‰, mean diameter of alluvium d_m=0.7 mm											
1.5	8.0	45.0	17.9	1.51	0.45	36					
1.7	10.0	55.0	26.0	1.56	0.63	96	6.7	1.0			
1.8	12.0	65.0	30.7	1.81	0.73	95	5.5	1.26			
1.9	14.0	80.0	33.0	2.03	0.92	52	6.2	0.93			
2.1	16.0	81.0	29.0	2.09	1.09	24	7.1	1.55			
River Terek, Stepnoye-Grebenskaya, flood 1982, S=0.4‰, d_m=0.5 mm											
0.5	1.2						2.5	1.9	0.15	0.05	
0.5	2.0	17.5	6.1				3.8	1.8	0.18	0.09	
0.5	2.8						4.8	1.9	0.23	0.09	
1.0	1.2	16.5	3.6				5.3	1.5	0.19	0.04	
1.0	2.0	15.4	5.7				4.9	2.2	0.24	0.12	
1.0	2.8	22.2	6.6				6.3	2.7	0.31	0.16	
1.0	3.6	23.6	9.1				6.9	2.7	0.37	0.20	
1.0	4.4	22.4	7.6				7.8	3.4	0.42	0.22	
1.5	1.2	11.9	2.7				6.3	1.4			
1.5	2.0	17.6	5.9				6.3	2.4	0.27	0.14	
1.5	2.8	20.3	7.4				7.6	2.9	0.42	0.23	
1.5	3.6	23.1	8.6				8.3	3.2	0.51	0.30	
1.5	4.4	30.2	8.1				8.2	3.7	0.45	0.25	
1.5	5.2	25.4	10.7				7.6	3.3	0.42	0.29	
1.5	6.0	37.8	16.7				9.2	4.3	0.53	0.33	
2.0	2.0	21.0	5.5				7.5	2.4	0.25	0.12	
2.0	2.8	18.1	4.6				8.3	2.8	0.33	0.18	
2.0	3.6	28.3	8.1				9.5	3.8	0.42	0.25	
2.0	4.4	29.3	11.8				8.4	3.3	0.43	0.25	
2.0	5.2	31.4	13.0				9.5	3.7	0.51	0.33	
2.0	6.0	41.6	15.0				10.9	4.7	0.53	0.39	
River Dniestr, Dubossary-mouth, low water 1984, S=0.11‰, d_m=0.3 mm											
0.7	2.5	8.8	3.8	0.40	0.18	979					
0.9	3.5	9.0	3.7	0.40	0.20	609					
1.0	4.5	8.2	3.8	0.33	0.20	274					
1.1	5.5	11.0	5.7	0.32	0.24	88					
1.1	6.5	10.2	4.0	0.21	0.11	63					
River Ob', Fominskoye, low water 1985, S=0.22‰, d_m=0.7 mm											
0.90	1.9						3.4		0.22		
0.99	2.7						2.6	1.12	0.16		179
1.04	5.3						2.2		0.12		

U m/s	D m	L_d	σ	h_d	σ	n	L_R	σ	h_R	σ	n
River Lena, Pokrovsk-Yakutsk, low water 1986-87, S=0.19‰, d_m=0.25 mm											
0.5	1.5								0.21	0.067	284
0.85	2.5						3.0	1.35	0.23	0.079	1265
1.1	3.5	12.5	5.2	0.53	0.24	31	2.9	1.46	0.24	0.081	1796
1.25	4.5						2.6	1.06	0.30	0.095	746
1.4	5.5								0.30	0.099	507
1.6	6.5								0.33	0.112	122
1.8	7.5	23.4	9.3	0.99	0.42	11			0.29	0.102	75
River Vychegda, Fedyakovo-mouth, low water 1997-98, S=0.08‰, d_m=0.3 mm											
0.7	2.0	13.0	6.0	0.20	0.08	144	3.0	1.2	0.1	0.044	440
1.1	1.2						3.8	1.8	0.16	0.06	623

Here Γ is gamma-function. Parameters α and β are correlated with megaripples mean length and standard deviation: $L_R=\alpha/\beta$; $\sigma=\sqrt{\alpha/\beta^2}$. Coefficient of variation $C_v=\sigma/L_R$ is constant and equals to 0.4 for ripples and megaripples formed in rivers and flumes in the wide range of hydraulic factors (Figure 2.6). Therefore distribution for their lengths is determined only by parameter β or by the mean length of these bed forms.

Wang and Chen (1980) offered Weibull distribution for description of the ripples height variability. The same distribution can be used for megaripples in the rivers (Figure 2.5b):

$$dp = \mu\,\lambda\,h^{\mu-1}\exp\left(-\lambda\,h^{\mu}\right)dh \qquad (2\text{-}2).$$

Parameters λ and μ are correlated with megaripple mean height and standard deviation: $h=\lambda^{-1/\mu}\Gamma\left(\frac{1+\mu}{\mu}\right)$; $\sigma_h=\lambda^{-1/\mu}\sqrt{\frac{2}{\mu}\Gamma\left(\frac{2}{\mu}\right)-\frac{1}{\mu^2}\left[\Gamma\left(\frac{1}{\mu}\right)\right]^2}$. As in the case of megaripple length, coefficient of variation σ/h is constant and equals to 0.5 (see Figure 2.6). Therefore the distribution is determined only by parameter λ or by mean height of these bed forms.

The mean length L_R of megaripples changes with hydraulic characteristics of flow, growing with the increase of flow depth and velocity (Figure 2.7). The type of relationship is the same as for the theoretical ultra-microforms (point N1 on Figure 1.1):

$$L_R = 5.4DFr \qquad (2\text{-}3).$$

Here $Fr=U/\sqrt{gD}$ is Froude number, U is the mean flow velocity, g is acceleration due to gravity and D is the flow depth.

Active megaripples in the rivers are slightly asymmetrical in long profile, with convex both upward and downward slopes. Distribution of asymmetry L_d/L_R (L_d is the length of the vertical projection of downward slope) can be described with gamma-distribution (Figure 2.5c) with modal value close to 0.25-0.3.

The linear analysis of initial perturbations does not allow obtaining a theoretical expression for megaripple height. Empirical data from the rivers (Table 2.2) and flumes (Yalin, Karakhan, 1979) can be described with the following formula:

$$\frac{h_R}{D} = a\frac{U-u_{cr}}{\sqrt{gD}}\exp\left(-b\frac{U-u_{cr}}{u_{cr}}\right) \qquad (2\text{-}4).$$

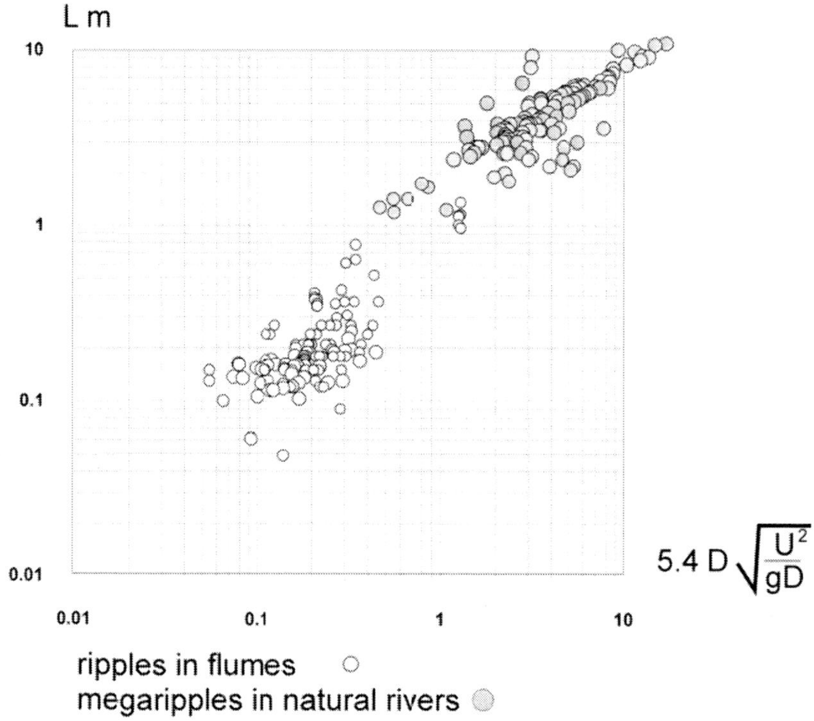

Figure 2.7. The relationship of ripple and megaripple lengths to hydraulic factors.

Figure 2.8. Bed forms (megaripples) on the surface of an alternating bar in the Alabuga River channel (Kirgizia), composed of the boulders. See geological compass 10 cm long for the scale.

Figure 2.9. Bed forms (megaripples) on the surface of an alternating bar in the Suor-Uyallakh River channel (Kular Ridge, Russia), composed of the schist debris. The band for the scale is 3 m long.

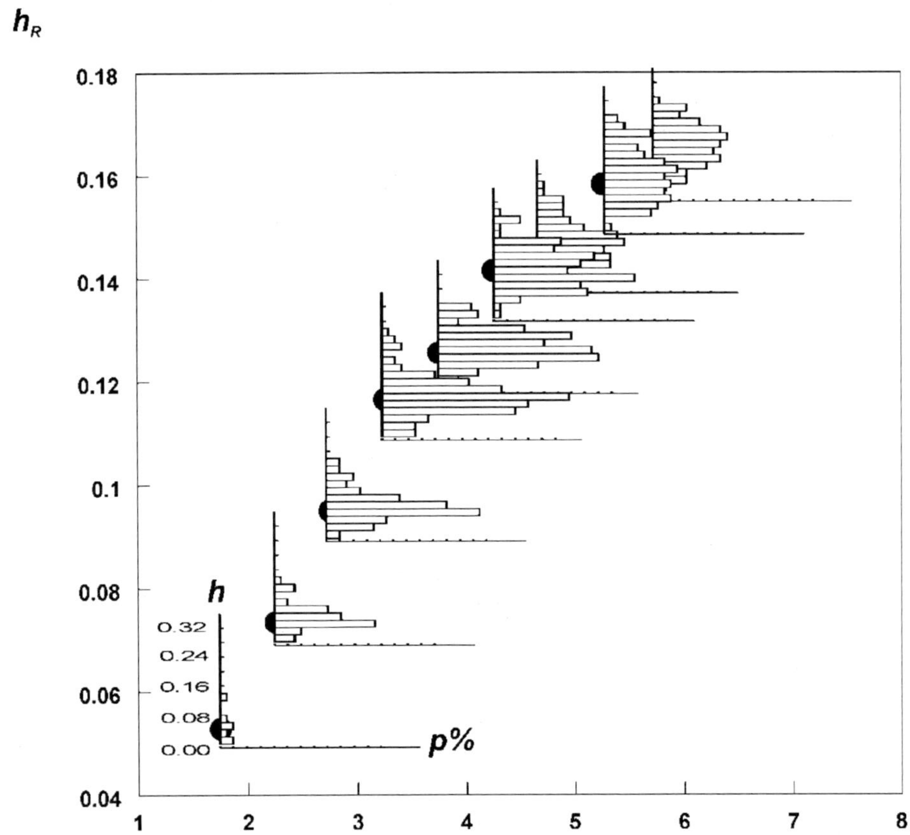

Figure 2.10. Stochastic relationship between megaripple heights and lengths. An example from the middle Lena River.

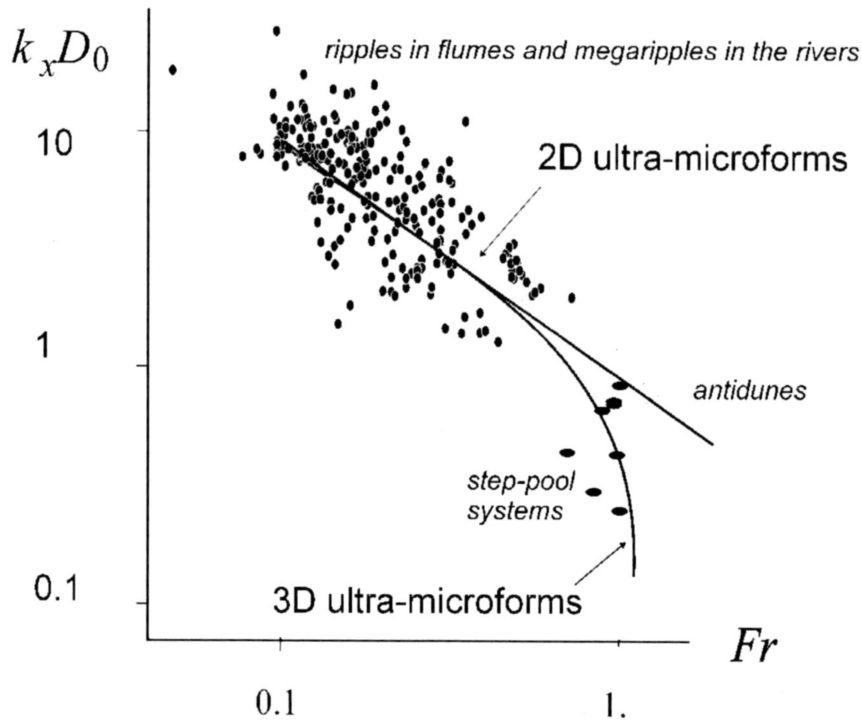

Figure 2.11. Correspondence of the theoretical ultramicroforms to the megaripples, antidunes and step-pool systems.

For the Lena River a=1.6 and b=0.5. When flow velocity is close to the critical velocity of the bed load transport initiation u_{cr}, the height of the megaripples is very small. However, megaripples are formed at the broad range of bed-load particle sizes, from silt and up to schist debris and boulders (Figure 2.8, 2-9).

Probability of megaripples steepness h_R/L and h_R/L_d is described with Weibull distribution (Figure 2.5d). The relationship between megaripple height and length is highly stochastic (Figure 2.10), but mean and standard deviations of height for each length category shows a clear tendency to increase with the length (Table 2.2).

Theoretical calculations show that the length of ultra-microforms (equal to megaripples) increases rapidly and nonlinearly when Froude numbers are high. Megaripples become relatively large when flow is close to critical. Under such conditions these bed forms change their morphology and dynamics and are called 'antidunes', as they have the size of dunes, but move upstream in a supercritical flow. Comparison of the observations in the large flumes (Guy et al., 1966) with the theoretical results (Figure 1.2) shows that antidunes belong to ultra-microform field and continue megaripples (not dunes) into the high Froude number field (Figure 2.11). So-called step-pool systems in highly kinetic mountain streams (Lenzi, 2001) also belong to the same field, which confirms the earlier findings of Whittaker and Jaeggi (1982).

Dunes

Dunes are bed forms with the length between ripples and bars (Nomenclature..., 1966). In this broad range we can often find several orders of bed forms, which are superimposed one on the other (see Figure 2.1).

The most common bed forms are dunes of the first order (dunes_1) (Table 2.2). In the lower Volga River channel they have the mean length of 45 m and height of 0.8 m during the low-water period (Table 3). Gamma-distribution (2-1) with parameters α = 3.51 and β = 0.089 fits in well with the dune length variability there. The Weibull distribution (2-2) with the parameters μ = 1.89, λ = 1.11 fits the height variability. The same distributions are characteristic for Niger River dunes_1 (Figure 2.12a, b) and for the dunes_1 at the rivers listed in Table 2.2.

Table 3. The hierarchical system of the lower Volga River alluvial relief (low water)

hierarchical level	alluvial feature type	mean wavelength/ standard deviation, m	mean height/ standard deviation, m	number of alluvial features
ultra-microforms	megaripples	~ 5	~ 0.1	
microforms	dune_1	43/31	0.8/0.5	624
	dune_2	140/83	1.0/0.67	149
mesoforms	dune_3	580/310	1.2/0.76	35
	sand bar_1	3100/830	up to 12	11
	sand bar_2. small islands	5900	up to 15	4
macroforms	large islands. meanders	13000	up to 20	
megaforms	parallel channels	>100000	>20	

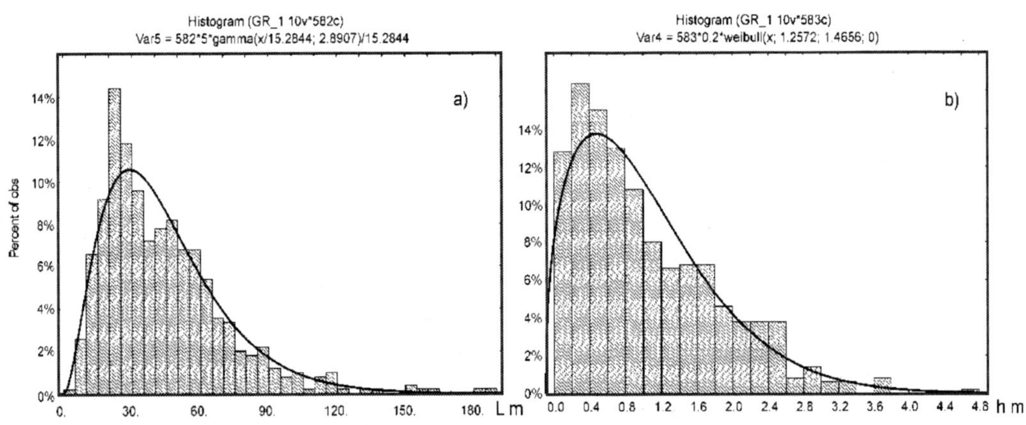

Figure 2.12. Histograms of active dune_1 geometrical variables with fitting functions (a – length; b – height) at the lower Niger River during the flood.

The above mentioned two types of probability density functions are preferred on the basement of some general considerations. Gamma distribution (the continuous version of Poisson distribution) is used to describe dune length PDF because sediment transport in a flow follows Poisson process (van Rijn, 1970; Sidorchuk et al., 2008). Weibull distribution is known as the distribution of maximums, therefore it is used to describe dune height PDF. Observation data analysis shows that a rather broad range of asymmetric PDFs fit the dune length and height empirical PDFs with high level of accuracy (Table 4).

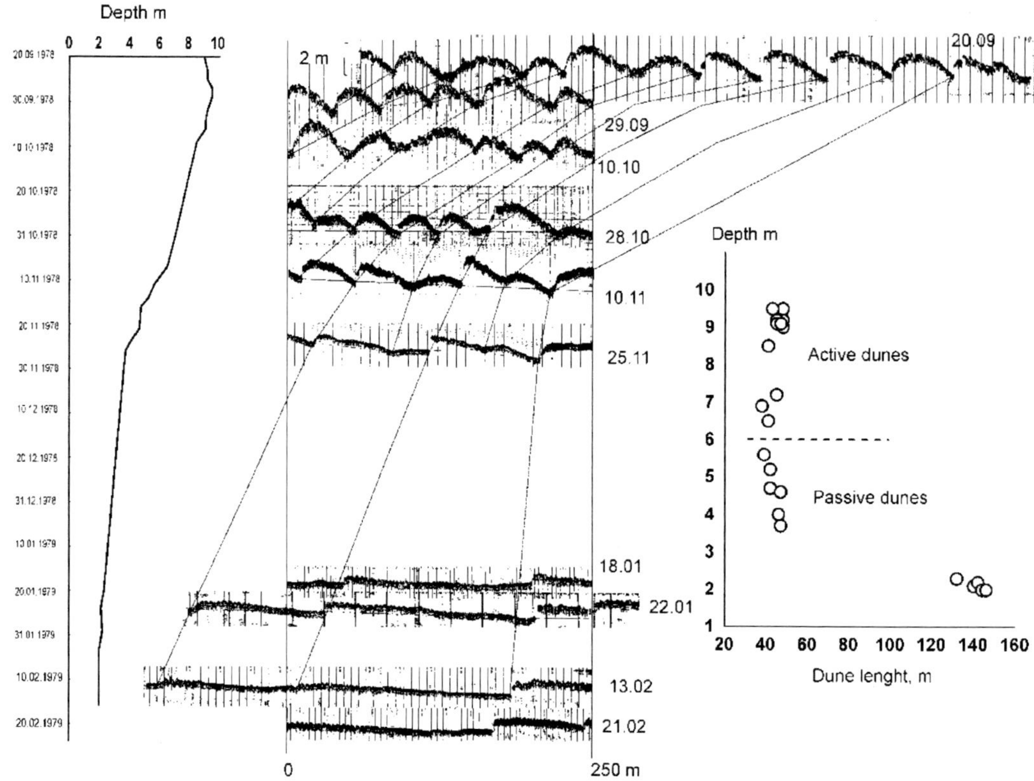

Figure 2.13. Transformation of dunes in the Niger River channel from active state during the flood to passive state during the low water period.

Table 4. Kolmogorov-Smirnov criteria for PDFs of the lower Volga River dunes

PDF type	L_d			h_d		
	dune_1	dune_2	dune_3	dune_1	dune_2	dune_3
extremal	0.066	**0.040**	0.093	**0.035**	0.112	0.102
Beta	-	-	-	-	-	-
Gamma	**0.057**	**0.050**	0.086	0.058	**0.085**	**0.069**
Weibull	0.064	0.076	**0.077**	**0.038**	0.101	**0.070**
Reley	0.068	0.095	0.085	0.100	0.186	0.182
normal	0.103	0.121	0.111	0.086	0.158	0.107
log-normal	**0.044**	0.075	**0.084**	0.111	**0.080**	0.099
exponential	*0.247*	*0.237*	*0.240*	*0.187*	*0.224*	0.165

First-order dunes in the river channels can occur at active and passive stages (Figure 2.13). Dunes at an active stage are conformed to flow velocity. Some attempts to reconstruct this initial flow field (Sidorchuk, 1996) show that flow velocity above dunes increases with the depth increase: depth-averaged velocity maximum is situated above dune trough and the minimum – above dune top. This wave-like flow field moves downward, and the dunes move accordingly.

Initial flow field is transformed by channel bed relief: dune growth leads to inversion of velocity distribution. Flow velocity above mature dunes increases with depth decrease: velocity maximum is situated above dune top and the minimum – above dune trough. At this stage velocity field is completely controlled by channel relief, and active dunes are changed into passive ones. Passive dunes are moved not by velocity field change, but by sediment particles and smaller bed forms movement along the upward dune slope and deposition at the downward slope.

At the Niger River near Adjaokuta during the flood maximum 27.09-7.10 1978 the depth of the channel near right bank was 9-10 m, mean flow velocity was 0.9-1.05 m/s. The channel bed was composed of sand with mean diameter 0.3-0.4 mm. Bed forms were represented by first order active dunes with the length 30-40 m, covered by megaripples with the length 3-5 m (Figure 2.13). Dunes had gentle (about 6°) slopes and were nearly isometric in longitudinal and cross sections. Their lengths were in correspondence with flow hydraulics. Flow field above active dunes was controlled by eddy-like structures (see Sidorchuk, 1996, Figure 3), and channel relief was generated by flow structure. With the water level drop during October – December, the relative dunes height increased and their shape become asymmetrical (triangular) in longitudinal section. The lengths of dunes increased due to aggregating of adjacent dunes. The variability of depth in cross-section along the dunes decreased and these bed forms become essentially 2D. Their lengths lost their correspondence with flow hydraulics. First order dunes were transformed into passive state. Flow field above passive dunes was controlled by depth distribution along the dunes (see Sidorchuk, 1996, Figure 3), channel relief generating the flow structure. Dynamics of such passive dunes was determined by the movement of smaller bed forms. Only in deep pools, where dunes relative height remained small, these bed forms were active even during the low water stage and remained symmetrical and 3D.

About 30% of the active first-order dunes at the channel bottom are nearly isometric in both longitudinal and cross section. Their shape can be described with elliptic functions, such as θ-functions. At the Niger River first-order dune asymmetry is described by a normal distribution with the mean equal to 1. The steepness of the downward slope of first-order dunes is well described by a gamma distribution with $\alpha = 1.95$ and $\beta = 37.8$. That corresponds to the mean steepness of 0.052 and standard deviation 0.037.

Passive first-order dunes are highly asymmetric, and their asymmetry increases with water level and flow depth decrease. Due to increase of flow velocity along the upward dune slope the top of a dune moves faster than the trough. A symmetric elliptic active dune is therefore transformed into a triangle passive one. These first-order passive dunes are well defined on the side-looking sonar images and are essentially 2D at the low water levels. Their tops form straight parallel lines across the channel, and their typical sizes can be measured using only one longitudinal bathymetric profile. As passive dunes at low water stages are observed more often than active dunes, there is a common opinion that dunes in general have an asymmetric triangle shape, and there is a reversed water flow at their downward steep slopes. Such flows are often modelled in flume experiments as the most typical for rivers. The same opinion prevails in an analysis of ancient alluvium structure. Cross-layer alluvial sequences are interpreted as formed by dune movement, which is true only for the passive dunes. Horizontal-layer alluvial sequences (see Reineck and Singh, 1975) are usually interpreted as formed by low energy flows or by flows with antidunes, but often these sequences were formed by high-energy flow with active dunes.

Second-order dunes, with the mean length 140 m and mean height 1.0 m, are also common in the lower Volga River at the low water-period (Table 2.3). Their morphology is described with the same distribution curves as that of the first-order dunes. The gamma-distribution for their lengths has parameters $\alpha = 3.09$ and $\beta = 0.0217$. An asymmetry of the second-order dunes profile is well approximated by the normal distribution. The Weibull distribution for their height has parameters $\mu = 1.60$ and $\lambda = 0.87$. The mean steepness of downward slope is 0.0234 with standard deviation 0.027, gamma distribution with $\alpha = 2.05$ and $\beta = 244.5$ fits well to the steepness distribution. The same distributions are characteristic for Niger River dunes_2 (Figure 2.14a, b).

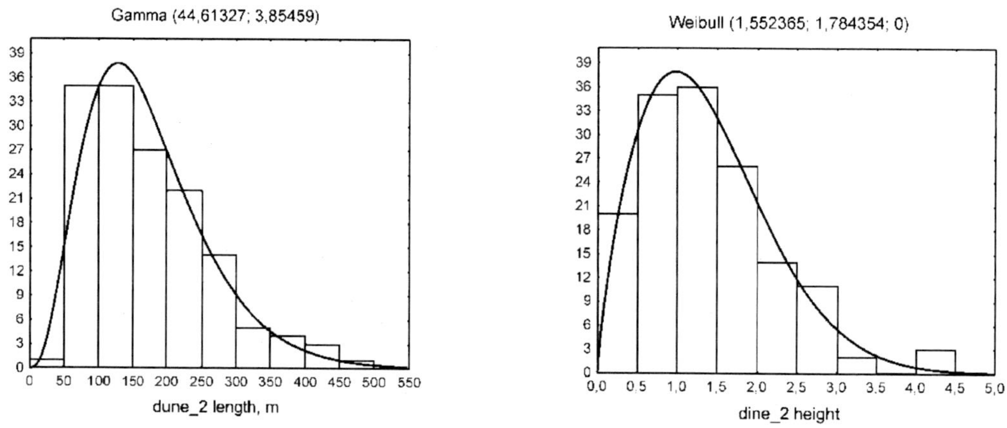

Figure 2.14. Histograms of active dune_2 geometrical variables with fitting functions (a – length; b – height) at the lower Niger River during the flood.

The relatively low steepness of the downward slope of second-order dunes makes them practically invisible on the side-looking sonar images, because steeper first-order dunes completely predominate on the image (Figure 2.15).

Dunes of the third order with the mean length 580 m and the mean height 1.2 m are less frequent features in the lower Volga River. Nevertheless, there are enough empirical data to obtain statistically reliable distribution curves of their geometry (Table 2.4). Gamma-distribution for the third-order dunes length has parameters $\alpha = 3.41$ and $\beta = 0.0058$. The Weibull distribution for the height has parameters $\mu = 1.49$ and $\lambda = 0.69$. The mean steepness of the downward slope is 0.0067 with a standard deviation 0.0048; a gamma distribution with $\alpha = 2.13$ and $\beta = 318.5$ fits well to the steepness measurements. The same distributions are characteristic for Niger River dunes_3 (Figure 2.16a, b).

An empirical formula for the active dune length correlation with the flow characteristics is similar to formula (2-1) for megaripples:

$$L_d = k_d D_0 Fr \quad (2\text{-}5).$$

The coefficient k_d is not constant and increases with the river size (Figure 2.17).

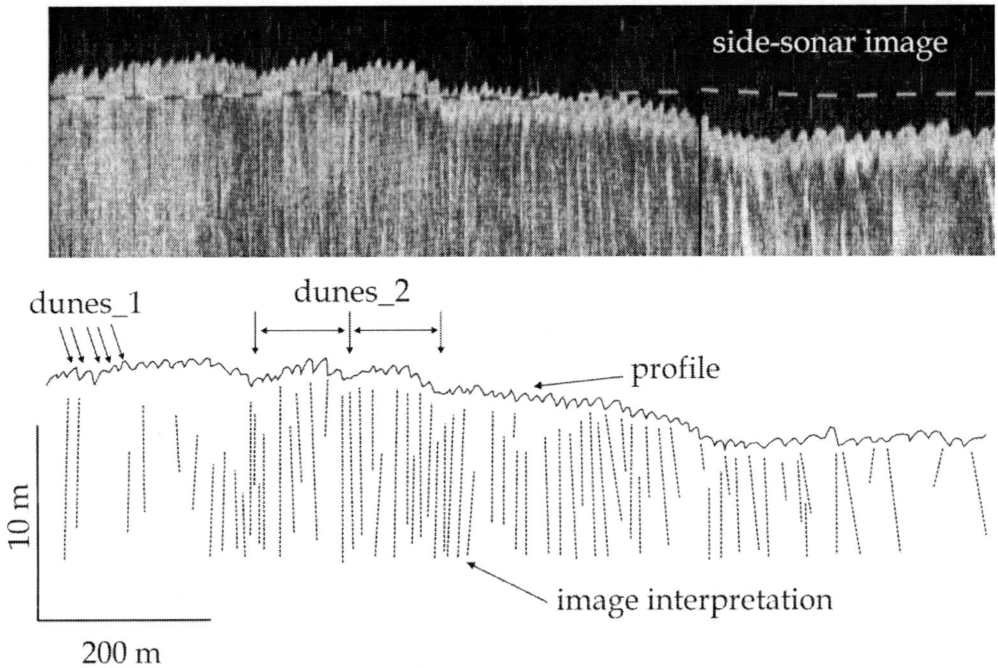

Figure 2.15. Dunes of the first and second order, profile and side-looking sonar image, the Volga River example (low water period).

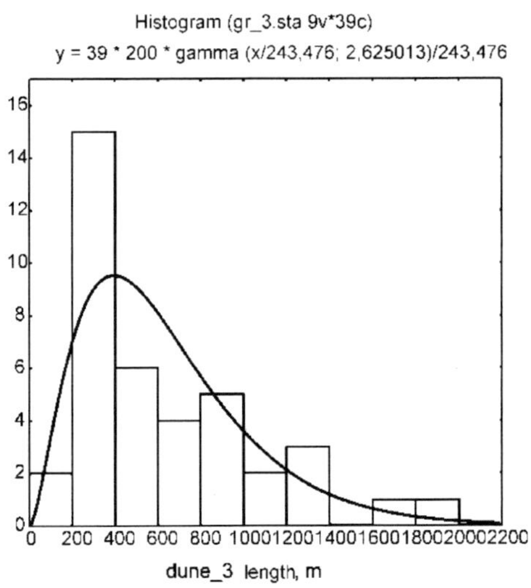

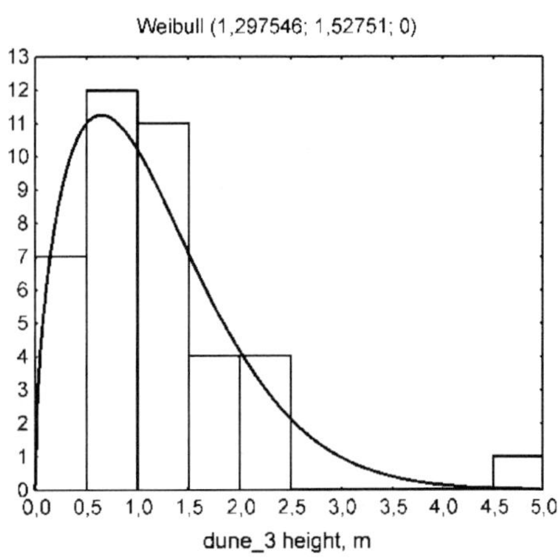

Figure 2.16. Histograms of dune_3 geometrical variables with fitting functions (a – length; b – height) at the lower Niger River during the flood.

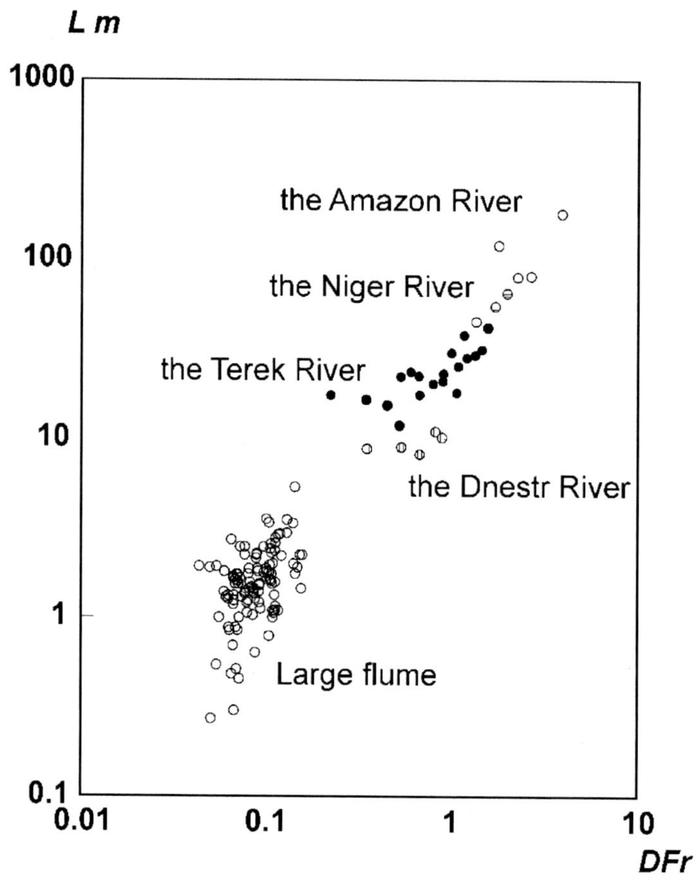

Figure 2.17. Relationship of dune lengths to hydraulic factors.

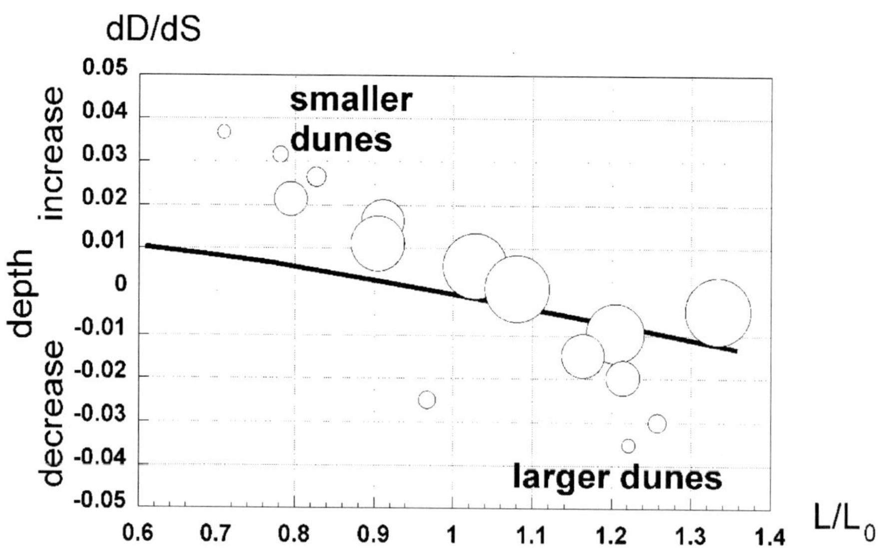

Figure 2.18. Relationship between active dunes_1 lengths and the flow depth longitudinal gradients in the Niger River during the 1978 flood. The size of circles is proportional to the number of dunes. The theoretical relationship is shown by the line.

The length of active dunes also depends on the channel bed general gradient. At the Niger River channel during the flood of 1978 (Figure 2.18) dunes_1 were longer, than on the flat bed (with the length L_0), at the channel segments with decreasing depth, and the ratio L/L_0 increased with depth negative longitudinal gradient increase. In opposite, dunes_1 were shorter, than on the flat bed, at the channel segments with increasing depth, and the ratio L/L_0 decreased with the depth positive longitudinal gradient increase. These observations are qualitatively consistent with the theoretical initial instability analysis for the straight main flow with the non-uniform longitudinal velocity and depth, but theoretical relationship is less expressive, than the empirical one (see the line for theoretical relationship at Figure 2.18).

The larger is a bed form, the shorter is the period of its active stage. Dunes of the second order were active in the Niger River channel during the year 1978 flood and became stable passive bed forms during the following low water period. Mostly these bed forms were completely reworked by moving 2D first-order dunes. Dunes of the third order remained passive bed forms during the year 1978 flood.

Dunes are the most common bed forms in large river channels with sandy alluvium (Amazon, Yamuna, Mississippi, Niger, Yenisei, Ob', and others). They are usually marked in the spectrum of natural channel bottom elevations with a well-defined local maximum (see Figure 2.2). Such a local maximum in the field of microforms is absent in the theoretical spectrum (see Figure 1.1). The discrepancy between the natural observations and theoretical results confirm the limitations of the linear approach to the initial-instability analysis.

Dunes of the first and second orders are usually bed (submerged) forms and correspond to the field of microforms. Dunes of the third order are also typically bed forms, but correspond to transitional field between microforms and mesoforms (Figure 2.19). In the Niger River and in the lower Volga River channel third-order dunes are mainly 3D bed forms, but during low water periods some of them can appear above the water surface. In that case they look like small bars and locally define the configuration of the water flow. In the channel of the upper Ob' River most of the alluvial features defined as third-order dunes (Figure 2.19) are represented by small bars and vegetation-covered islands even at bankfull water level. These small bars and islands determine very complicated isle-braided channel pattern of this river.

Bars

Alternating bars and braid bars often form the internal structure of the channel. These alluvial features are not covered or only partly covered by vegetation and are submerged during average floods. Alternating bars and braids cause additional sinuosity of the low water flow. There are two orders of bars in the lower Volga River channel. Bars of the first order have the mean length of 3100 m with standard deviation 830 m. A gamma-distribution with $\alpha = 14.46$ and $\beta = 0.0046$ can be used for their length description. The height of the first-order bars ranges from 2–3 to 11–12 m. Second-order bars there have the mean length of 5900 m with standard deviation 1410 m, their height ranges being from 4–5 to 12–15 m. A gamma-distribution with $\alpha = 20.5$ and $\beta = 0.0035$ can be used to describe their lengths. Their movement along the islands causes the alteration of resistance to flow in the main and secondary channels, which causes the quasi-periodic decreases and increases of the discharge in the main channel.

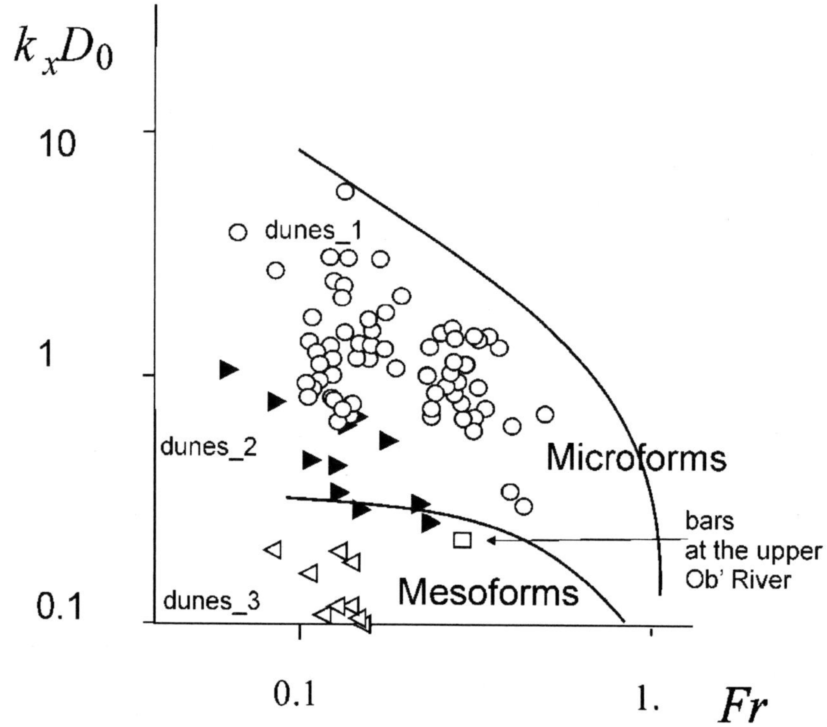

Figure 2.19. Correspondence of the theoretical microforms and mesoforms to dunes of different order observed in flumes and natural rivers.

On the Niger River within the 20-km long section between the Boqua and Shuter Islands, alternating second-order bars in the main channel, as measured between adjacent riffles, are ~4000-5000 m long. Each of these large bars consists of 2-3 smaller first-order bars with the length of 1000-1500 m (Figure 2.20).

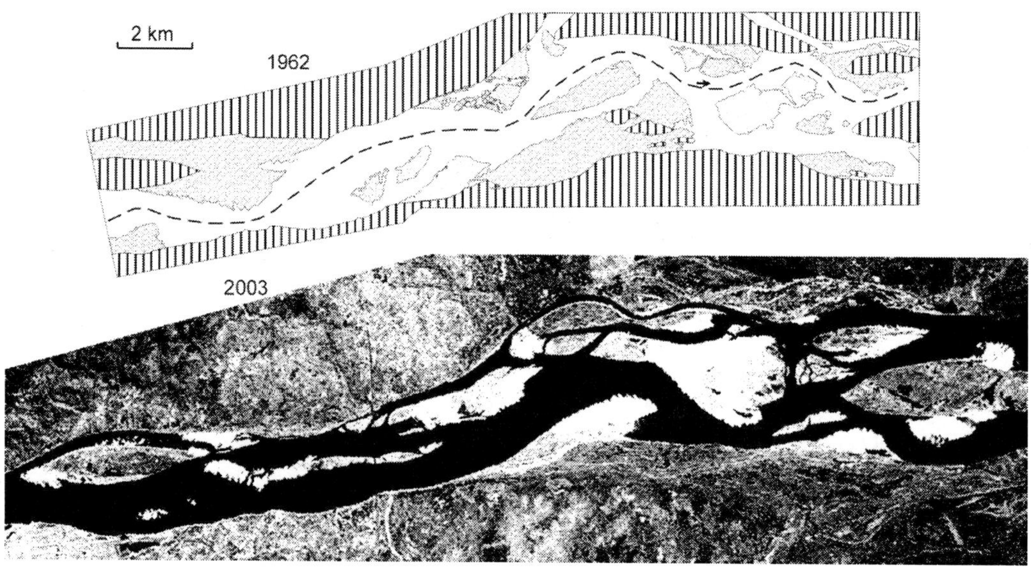

Figure 2.20. Bars and islands at the Niger River near Adjaokuta.

At the diagram of theoretical unstable alluvial features (see Figure 1.2), bars belong to the fields of mesoforms and macroforms. Their lengths increase with the main channel depth D_0 and Chezy coefficient C_0 value, but decrease with the growth of Froude number (Figure 2.21).

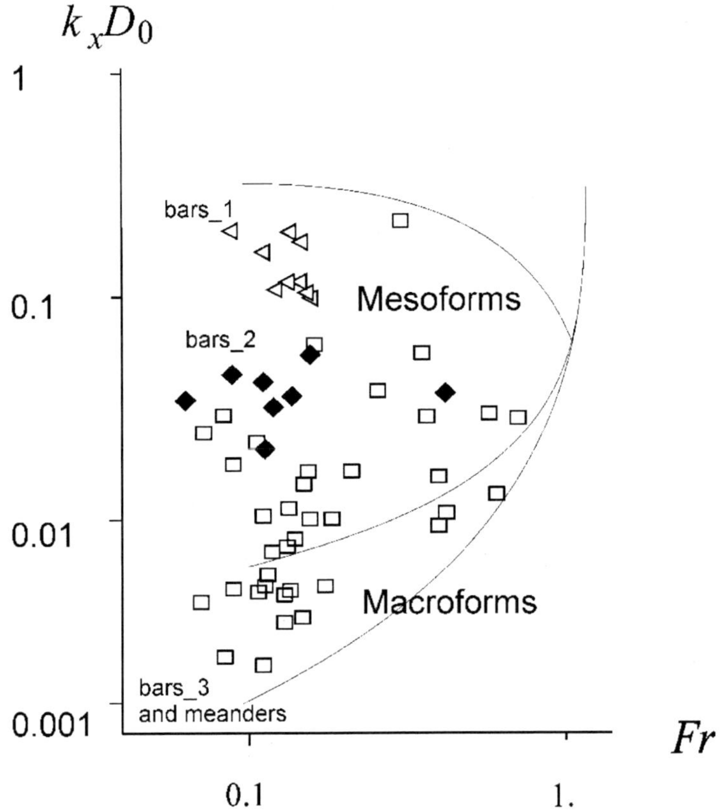

Figure 2.21. Correspondence of the theoretical mesoforms and macroforms to bars of different order (including the stabilized ones) observed in natural rivers.

The boundary between mesoforms and macroforms in the theoretical spectrum (Figure 1.1) is not clear. This raise some classification problems, because the lengths of small mesoforms and of microforms (which can be both dunes and bars) increase with Froude number, and the lengths of large mesoforms (which can also be both dunes and bars) decreases with Froude number (see Figure 1.2).

Large bars are the most stable alluvial features in the hierarchy of river channel forms. They can be active only during the most powerful floods, being in passive state the most part of the hydrological year. During the passive stage bars are triangular and asymmetrical in the long profile, with a steep downward slope. Their dynamics is controlled by movement of the smaller alluvial features over them. During the low water stage, the tops of bars are usually above the water level.

The large bars_2 and bars_3, islands and channel meanders usually correspond to the field of macroforms. As the dunes, these features are most common in river channel morphology. Well-defined maxima in the spectrum of natural channel bottom elevations correspond to these forms (see Figure 2.2). A local maximum of the rate of amplitudes growth also exists in the field of macroforms in the theoretical spectrum. This maximum represents the length and width of the most probable macroforms, which are described with formulas (1-28) and (1-29).

The wavelengths of observed large bars and meanders are often smaller than the wavelengths of the maximums in the theoretical spectrum. The main reason for this phenomenon is the influence of non-linear effects of channel development. The rate of growth of the amplitudes of the long bottom waves is related not only to the initial instability, which is described by the theoretical spectrum, but also to these effects, which become very important with the amplitude increase.

When the width of the long unstable wave is less than two widths of the channel, the influence of this alluvial feature on the bank erosion is small. The channel banks are straight, and braid bars (or islands) are developed within these banks (see Figure 1.3c). At such multiple-branched channels the observed wavelength of the large bars and islands is close to its theoretical value from the formula (1-28).

When the width of the long unstable wave (the initial bar) is more than two widths of the channel (Figure 1.3b), bank erosion takes place around this bar. The alluvial feature often stops its down-channel movement, being stabilized by vegetation and fine floodplain alluvium, and forming a meander. The rate of increase of the deflection angle Θ of this meander decreases with increase of its length. For the case of sine-generated curve this rate can be calculated with the formula:

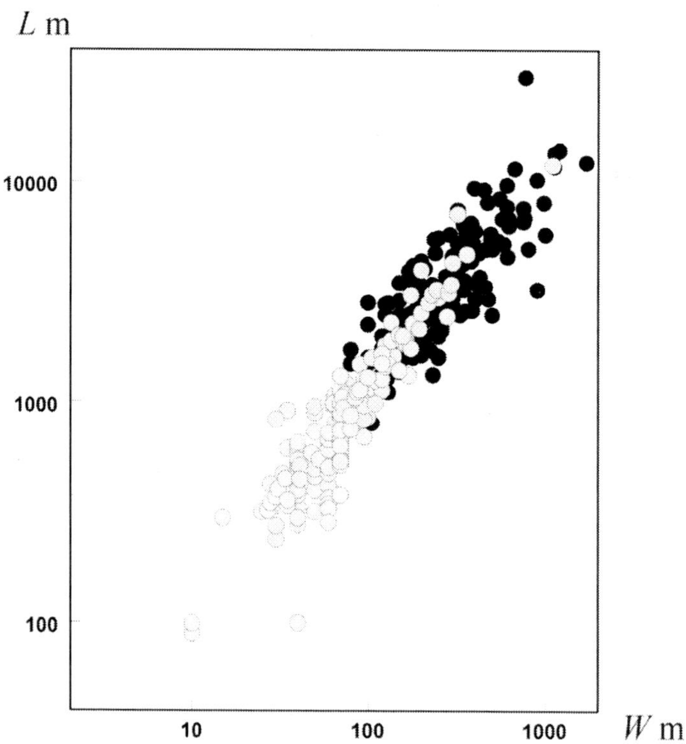

Figure 2.22. Relationship between channel bankfull width and meander wavelength for the rivers of the northern Eurasia. Grey circles are for the recent rivers, black circles are for the Late Glacial palaeochannels.

$$\frac{d\Theta}{dt} = \frac{4q_b}{L_m H_b} \frac{I_0(\Theta)}{dH_0/d\Theta} \quad (2\text{-}6).$$

Therefore the undulation with the lateral wavelength $L_Y \sim 2W_0$ grows faster, than the longer unstable undulations, and this is the most probable size of the mature meander. The most common value of length/width ratio for macroforms L_x/L_Y is about four (formula 2-29), so the length of mature meanders L_m must be about eight widths of the channel W_0. That a common value of meander length/ channel width ratio is 8-10 is confirmed by numerous data (Figure 2.22) (see Leopold and Wolman, 1960).

As the longer unstable undulations have lower rate of deflection angle growth, transformation of more active shorter mature meanders leads to increase of their amplitudes and deflection angles up to critical values, when these active meanders become straitened or cut off the main channel. These avulsions of active meanders occur several times in the course of the long-term river channel evolution. Transformation of the longer and less active undulations is slow, but steady, and after a long period of channel transformation these undulations become quite pronounced in the river channel geometry. This mechanics can be the cause of compound meander formation (Figure 2.23).

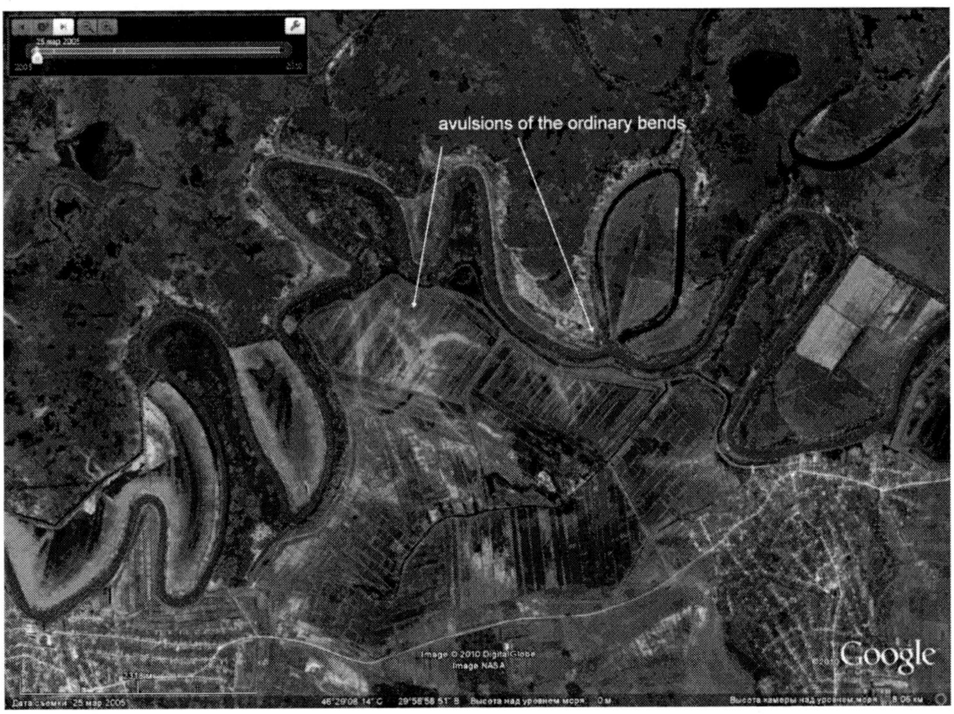

Figure 2.23 A compound meander of the lower Dniester River. See the numerous avulsions of the ordinary bends.

Parallel Branches

The phenomenon of parallel branches in a natural river channel was first described by Berkovich et al. (1989). It occurs quite often in wide and relatively shallow river channels, when two branches of the same river are following parallel courses for a long distance, usually near the opposite sides of the valley bottom. These branches are divided by a chain of islands and are usually connected only by short lateral channels. This phenomenon is known at the lower reaches of Lena, Yenisei, Ob', Northern Dvina and Volga rivers. Presumably, a river delta also shows examples of this phenomenon. Parallel branches may represent a natural analogue of the theoretical megaforms. There had been relatively little investigations of very long and narrow forms of channel relief. Coupled lateral vortices with the horizontal axis along the channel were studied by Losievsky (1934). Nezu and Nakagawa (1989) described the longitudinal narrow strips of ridges and troughs, formed by such vortices in a flume.

Correlation between Alluvial Features belonging to Different Levels of Hierarchy

The ratio between the mean lengths of alluvial features, which belong to adjacent levels of hierarchy, is relatively constant at different rivers. Its mean value is 3.7 for megaripples, dunes and small bars and 3.2 for large bars (see Table 2.1), with the mean value 3.5 for the whole hierarchical complex.

Correlation between the lengths of individual alluvial features at different hierarchical levels is low. For example, the lengths of 15 dunes_2, 60 dunes_1 and 630 megaripples were measured within 4.1 km profile of the Niger River bed between Geregu and Adjaokuta obtained on Nov. 7, 1978. The lengths of dunes_1 were averaged for each dune_2, and the lengths of megaripples were averaged for each dune_1. The correlation between these values is very low: it is 0.2 between dune_2 and dune_1 and between dune_2 and megaripple lengths, and 0.3 between dune_1 and megaripple lengths. Similar results were obtained for the hierarchy of channel relief at the Yenisei River near Dudinka, the Oka River near Puschino and the Terek River near Stepnoye (Table 5). These examples show a relative independence of individual alluvial forms within their hierarchy, along with rather close relationships of their associations.

The variability of morphological parameters of alluvial features decreases from lower to higher hierarchical levels. The variation coefficient C_v for the lengths of alluvial forms (see Table 2.1) is 0.4 for megaripples and dunes_1, 0.3-0.35 for dunes_2, 0.25-0.3 for dunes_3 and bars_1, and 0.16-0.31 for bars_2 and meanders. Presumably, the reason for this tendency can be a stabilising influence of the river channel width. This influence generally increases with increasing relative size of an alluvial feature. In some cases, such as at the Niger River, this tendency can be broken for the largest channel forms, and the variability of meander lengths can become greater. It is brought about by a variable lithology of the river banks, which represents an additional factor of alluvial forms variability.

Alluvial features show a clear tendency to a decrease of steepness (h/L) with an increase of the length. Despite a very low correlation between heights and lengths of individual megaripples, dunes and bars, the mean steepness, calculated for their associations, decreases with alluvial feature growth (Figure 2.24).

Table 5. Matrix of correlation coefficients for the lengths of alluvial features in the hierarchy

alluvial feature	index	1	2	3	4	5	6	7	8
The Niger River									
meander	1	1.0	0.5	-0.2	-0.1	-0.1	-0.2		
bar_3	2	0.5	1.0	0.6	0.1	-0.3	-0.2		
bar_2	3	-0.2	0.6	1.0	0.6	-0.3	-0.2		
braid bar_1	4	-0.1	0.1	0.6	1.0	0.6	-0.2		
dune_3	5	-0.1	-0.3	-0.3	0.6	1.0			
dune_2	6	-0.2	-0.2	-0.2	-0.2		1.0		
dune_1	7						0.2	0.2	0.2
megaripple	8						0.2	1.0	0.3
The Enisey River									
dune_2	6						1.0	0.4	0.5
dune_1	7						0.4	1.0	0.5
megaripple	8						0.5	0.5	1.0
The Oka River									
bar_2	3			1.0		0.72	0.43	0.09	
dune_3	5			0.72		1.0	0.68	0.35	
dune_2	6			0.43		0.68	1.0	0.43	
dune_1	7			0.09		0.35	0.43	1.0	
The Terek River									
bar_2	3			1.0	-0.04	0.45	0.59		
bar_1	4			-0.04	1.0	-0.27	-0.38		
dune_3	5			0.45	-0.27	1.0	0.61		
dune_2	6			0.59	-0.38	0.61	1.0		

A complexity of alluvial relief can be defined as the number N of alluvial features, which composes the hierarchical alluvial relief. The smallest alluvial features in the natural rivers which depend on flow hydraulics (megaripples) are characterised by the lengths $L_R = 5.4DFr$, the largest ones (macroforms) have the lengths $L_x = 6.28D\left(\frac{C_0^2}{2g}\right)Fr^{-1}$ in the multiple-branched channels and L_x=8-10W_0 in the meandering channels. An average ratio between the mean lengths of any two adjacent alluvial features in the hierarchy is Rt~3.5. Therefore, the number of alluvial features between megaripples and macroforms (meanders, large islands or bars) can be calculated as

$$N = \frac{\ln(L_x/L_R)}{\ln Rt} \tag{2-7}$$

This formula can be transformed into

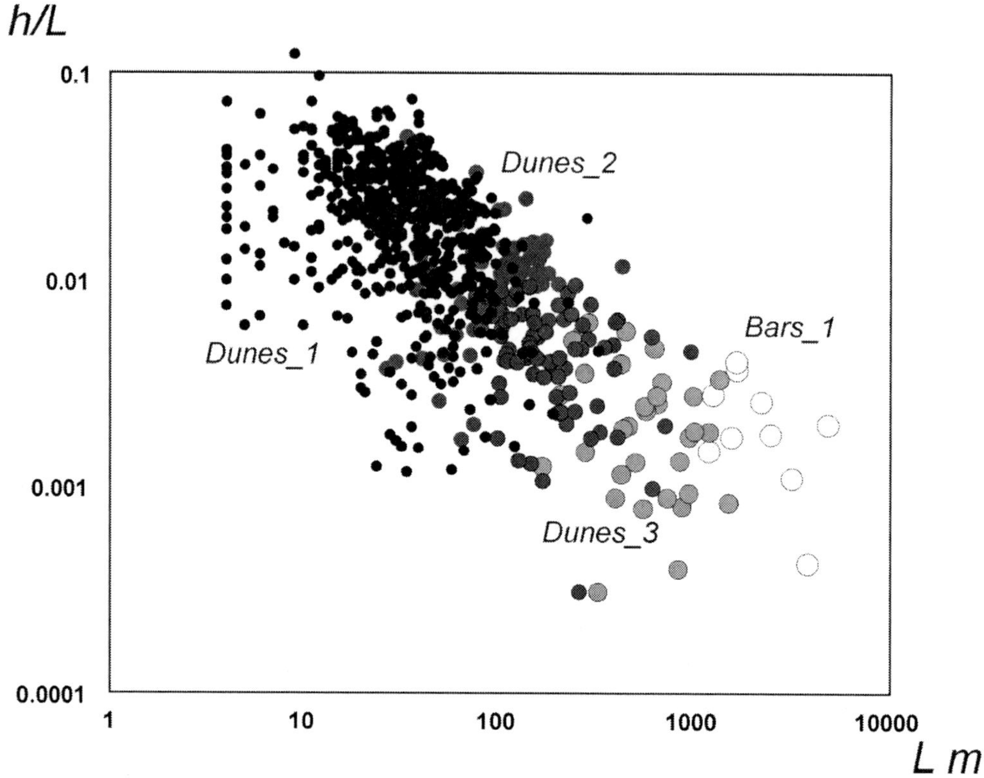

Figure 2.24. Correlation between steepness and length of individual dunes and bars during the low water period, an example from the Volga River.

$$N \approx 0.8 \ln\left(0.58 \frac{C_0^2}{g} Fr^{-2}\right) \approx 0.8 \ln\left(0.58 S^{-1}\right) \tag{2-8}$$

for multiple-branched channels and into

$$N \approx 0.8 \ln\left(1.85 \frac{W}{D_0} Fr^{-1}\right) \approx 0.8 \ln\left(1.85 \frac{\sqrt{g}}{C_0} \frac{W}{D_0} S^{-0.5}\right) \tag{2-9}$$

for single-channel meandering rivers.

Complexity of alluvial relief does not depend directly on the river size: it decreases in the mountainous rivers with high water surface slope S and increases in lowland rivers with low slope. Large rivers are mostly lowland, their water surface slope being 0.01-0.1 m/km. According to formula (2-8), an alluvial relief complexity for these channels is from 7 to 9. In general, this is consistent with the data in Table 2.1.

A theoretical classification of alluvial features shown in Figure 1.2 can be supplemented with the empirical data from Tables 2.1, 2.2, and 2.3 and similar information from various publications (Figure 2.25). This semi-empirical diagram gives an opportunity to estimate a typical hierarchy of alluvial features under given hydraulic channel-forming characteristics (channel resistance, flow width, depth, and velocity). It is also possible to classify alluvial features in a given channel and to compare alluvial features in different rivers. This diagram is of the main importance for the river channel pattern estimation.

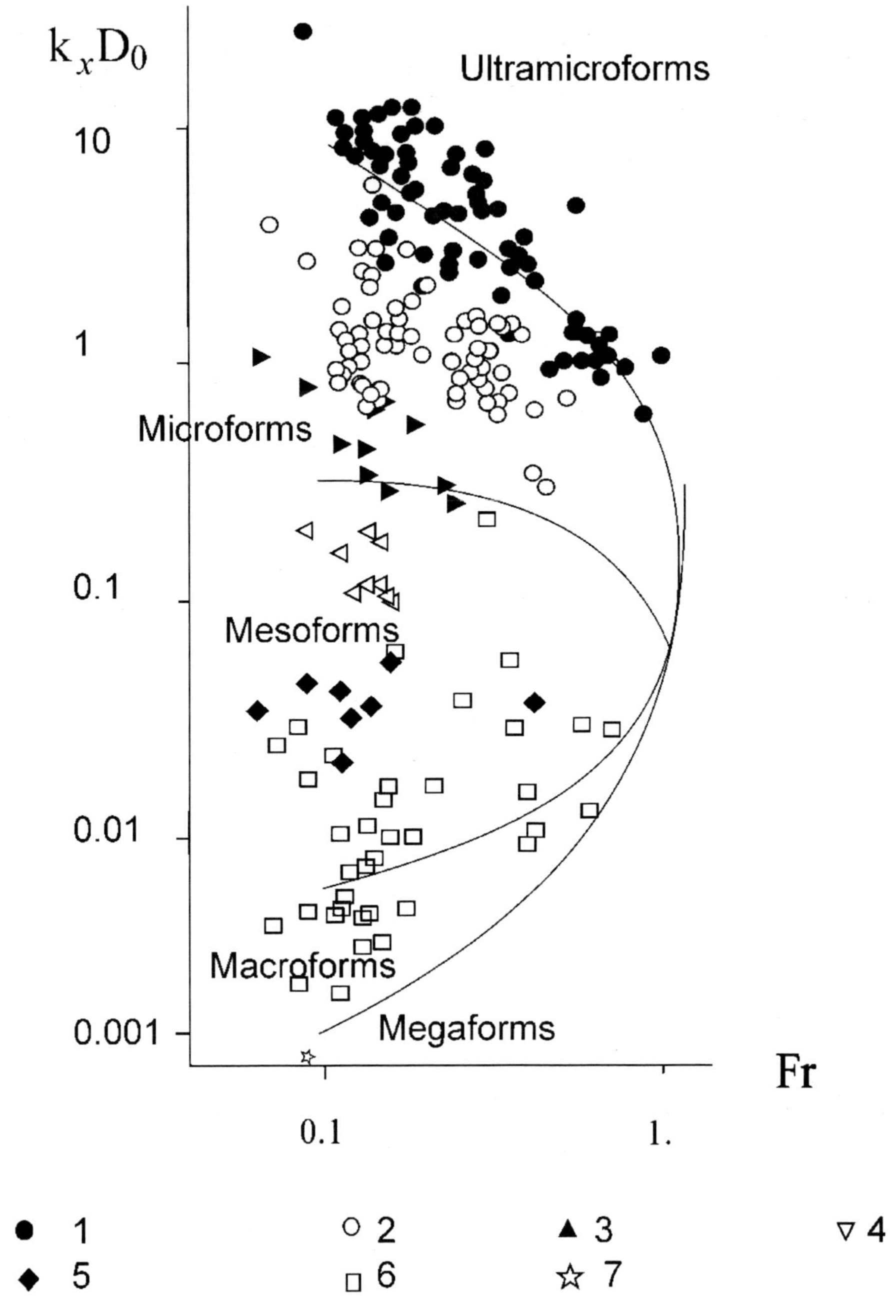

Figure 2.25. Correspondence of the bed and alluvial forms observed in flumes and natural rivers to the theoretical types of river bed undulations. 1 – megaripples, ripples and antidunes; 2 – dunes_1; 3 – dunes_2; 4 – dunes_3; 5 – bars_1; 6 – bars_2, bars_3 and meanders; 7 – parallel braids.

PART 3

Alluvial River Channel Patterns in Hierarchical complexes

The main characteristics of river channel geometry at the plane surface are controlled by the channel width and by the lateral wavelength of the largest alluvial features (see Figure 1.3). This principal geometry is complicated by smaller alluvial features of various hierarchical levels. The hierarchy of alluvial relief makes a rather simple principle configuration of a river channel very compound, but still predictable if the main parameters of the flow hydraulics are known.

The lateral wavelength of the largest alluvial feature is the function of flow velocity, depth and channel resistance (formula 1-29). If the width/lateral wavelength ratio is ≤0.5

$$\frac{W}{L_y} = \frac{1.5}{C_0}\frac{W}{D_0}\sqrt{\gamma_0}Fr^{1.1} \leq 0.5 \tag{3-1}$$

the most probable river geometry is a single channel, either straight with alternating or lingoid bars, or meandering. The characteristic wavelength of these bars or meanders is more than or (with high probability) equal to

$$L_x \geq 2W\sqrt{\frac{C_0^2\gamma_0}{g}} \approx 0.2C_0W\,. \tag{3-2}$$

Additional $N - 1$ hierarchical levels of bars and bed forms (with N from formula 2-9) complicate the channel shape. Their smaller wavelengths can be estimated from Figure 2.25, and their heights from Figure 2.24. These smaller bars can divide single-channel river into several branches of different width and depth.

The relationship between the channel width and flow hydrology can be estimated according to the regime equations approach. 450 sections of meandering rivers with sand alluvium located in the Russian Plain, West and East Siberia (Figure 3.1) were used to establish the following correlation between the bankfull channel width (W) and the mean maximum discharge (Q_{max}):

$$W = 3.7Q_{\max}^{0.56} \tag{3-3}$$

This simple relationship is rather general, as independent data of Xu (2002) led to a very similar formula.

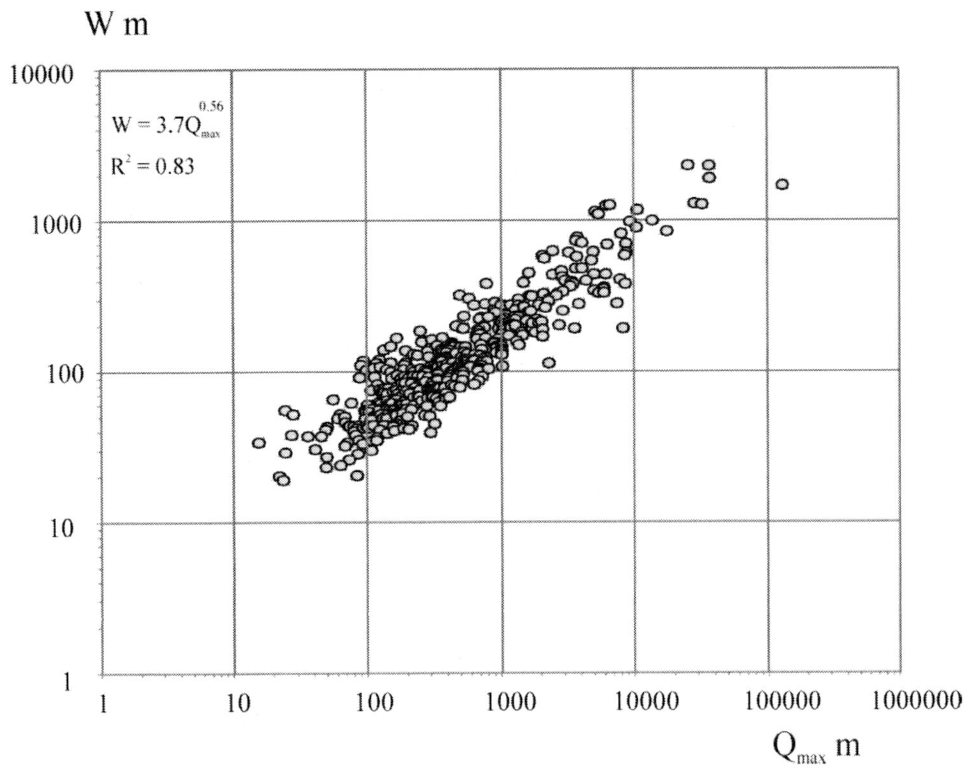

Figure 3.1. Relationship between channel bankfull width and the discharge of mean annual flood.

If the width/lateral wavelength ratio is between 0.5 and 1.0, the most probable river geometry is a channel with two main branches. Such a channel can be straight with alternating bars, separated from the banks by chutes, or straight with a chain of braid bars along the middle line of the channel. It can also be meandering with the chutes or braided by the chain of islands. The characteristic wavelength of these bars, islands or meanders is close to its theoretical value estimated with formula 1-28. Additional $N - 1$ levels of smaller bars and bed forms (with N from formula 2-8) complicate the shape of such channel. The wavelengths of all alluvial features can be estimated from Figure 2.25, and heights – from Figure 2.24. Smaller bars can divide the main branches of the channel into multiple ones.

If the width/lateral wavelength ratio is between 1.0 and 1.5, the most probable river geometry is a channel with three branches, and so on. The probability of such multiple-branched channel formation is higher for the mountainous rivers. In the mountains, the river valley bottom can be wider in the intra-mountain basins and lateral wavelengths of the large bars smaller due to a high water surface slope. For the lowland rivers the most favourable conditions of multiple-branched channel formation occur in their lower reaches, where the valley bottom width is often large enough.

Velikanov (1954) formulated the principal of the restriction of natural river channel complexes, bearing in mind that a unique combination of flow hydraulic characteristics determines a unique combination of channel morphological characteristics. Velikanov presumed that this principle can be a basement of a morphologic classification of the river channel patterns. All the arguments discussed above are within the mainstream of this principle.

The main limitation of this principle is the broad range of additional natural factors which have influence on the river channel patterns. These factors are the climate and landscape in the river basin; geological structure and texture of the river valley; the history of the river valley evolution.

The climate and landscape determine the main characteristics of the river flow: the mean discharge and hydrologic regimen, an average longitudinal slope of the river valley. Beyond these hydrologic characteristics climate, landscape and geology determine the grain size and amount of sediment load in the river, composition of the channel alluvial banks, the rate of floodplain accretion, floodplain vegetation types and successions.

Discharge and the river valley average slope were believed to be the main characteristics, which determine the channel pattern (Leopold, Wolman, 1957). Theoretical analysis shows that the relationships are more complicated. Dynamic system 'flow – river channel' is a dissipative system. For a river segment with long-term stable hydrologic regimen and channel morphology, an energy dissipation can be expressed by external factors – water discharge Q and average river valley slope S_0, as well as by river channel sinuosity P.

$$\varepsilon = \frac{\rho g Q S_0}{P} \tag{3-4}$$

The same characteristic can be expressed with internal flow parameters: velocity U, channel width W and channel resistance λ:

$$\varepsilon = \rho \frac{\lambda}{2} U^3 W \tag{3-5}$$

Therefore,

$$QS_0 = \frac{U^3 D}{g} \lambda \frac{W}{D} P \tag{3-6}$$

Here

$$Q = UDW \tag{3-7}$$

Simple transformations show that formula (3-6) with (3-7) is equal to the well-known empirical Chezy formula for flow velocity estimation. All external factors of the 'flow – river channel' system are grouped in the left part of (3-6). These factors are constant for a long-term and controlled by climate and landscape. Therefore, a given river stretch is characterized with the unique QS_0 value. The factors in the right part of (3-6) are internal for the 'flow – river channel' system. These are flow velocity U, flow depth D, the parameter $\lambda W/D$, which characterizes the lateral flow stability and, therefore, channel pattern, and sinuosity P, which also characterizes channel pattern.

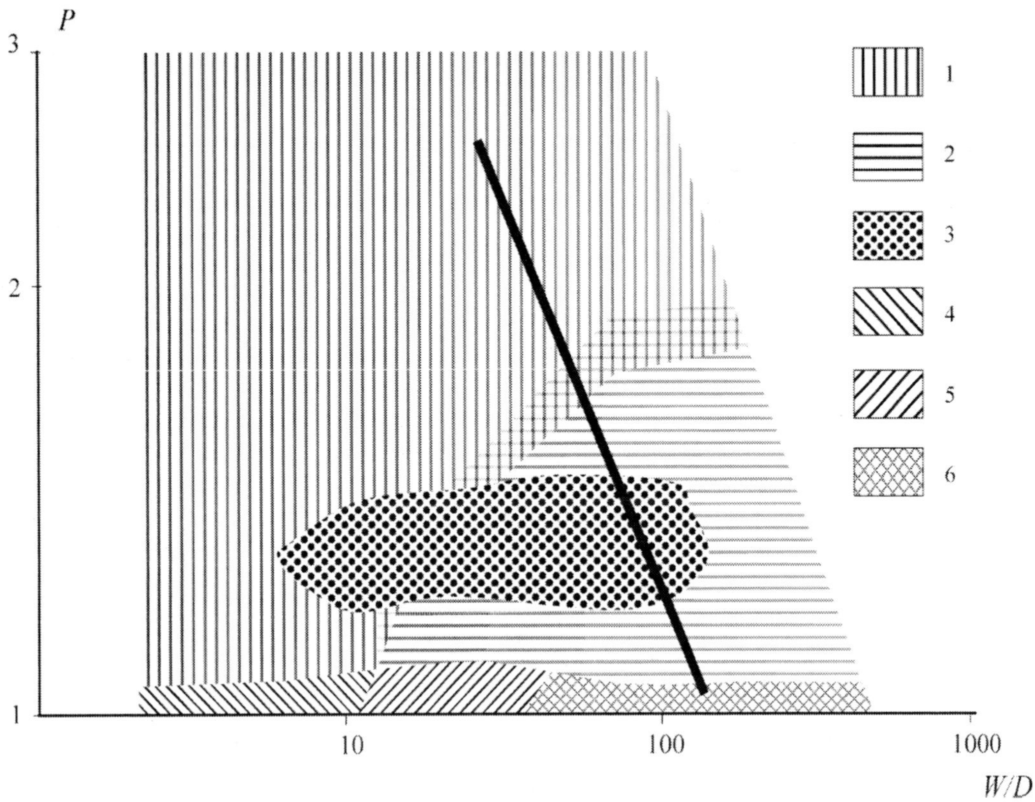

Figure 3.2. Positioning of the lower Vychegda River hydraulic characteristics (the black line) on the channel sinuosity *P* – relative width *W/D* diagram. Fields of the main river channel types: 1 – single-channel meandering; 2 – multiple channelled with islands and bars; 3 – meandering with chutes; 4 – single channel straight with linguoid bars; 5 – single channel straight with alternating bars; 6 – straight with braid bars.

It is obvious that any unique QS_0 value can be composed with multiple combinations of the specified internal factors. For example, let us take all flow parameters for the Vychegda River during the flood period: Q=7900 m^3/s; S_0=0.00008; U=1.2 m/s; D=7 m; λ=0.0029. For sinuosity in the range 1.0 - 3.0 the width/depth ratio calculated with formula (3-6) is in the range 174 – 58. These values suit all possible river channel types (Figure 3.2). The real lower Vychegda River has now sinuous-isle-braided channel with P=1.1 and W/D=160. In practice this means that simple *Q-S* diagram of Leopold and Wolman cannot be used for river channel pattern estimation.

The theory shows that the same hydraulic parameters, which are related to discharge and slope in formula (3-6), being organized in the other way (formula 3-1), make it possible to find the boundary between one-channel and multiple-channel rivers.

The climate and landscape determine the river hydrologic regimen, which governs the alluvial feature hierarchy changes through time. It is obvious that each discharge in the river channel has some influence on the river channel relief, but the degree of this influence varies for different discharges and alluvial features. Makkaveev (1955) stated that according to the general opinion, the magnitude of transformation TM_i of channel relief by a given discharge Q_i is proportional to the total amount of bed load transport $Q_{si}T_Q$ during the time span of this discharge T_Q. Makkaveev used *TM/Q* diagram to find the "effective" or "channel-forming" discharge, for which *TM* is maximum. Chalov (1979) specified four main types of *TM/Q* diagrams for the rivers of the former USSR (Figure 3.3), which differ from each other by the number of *TM* maximums and their relation to bankfull discharge. According to this author, rivers with *TM* diagrams of types A and B are characterized by complicated and active relief of the floodplain transformed by high discharges much greater than bankfull ones. For the same types of *TM/Q* diagrams, one (type A) or two (type B) maximums of *TM* are also typical, formed by discharges below the bankfull ones. These discharges produce active transformation of alluvial features within the channels even during the low water stages. Rivers with *TM* diagrams of types C and D are characterized with more simple relief of the floodplains, as discharges above the bankfull ones are less effective there. Alluvial relief transformations are concentrated within the channel banks when bankfull discharge (type C) or even lower (below bankfull) discharges (type D) are the most effective.

According to these findings, river channel patterns must be estimated with formulas (3-1), (3-2), (1-28) and (1-29), separately for different effective discharges. At the rivers with *TM* diagrams of types A and B, channel width during a high flood (at the first channel-forming discharge) is close to the width of the river valley bottom. To calculate the mean flow depth, velocity and resistance at this stage both the channel and floodplain characteristics should be integrated. Usually such rivers are multiple-branched, often of anastomosed pattern. Each of these multiple branches is effectively transformed by the second channel-forming discharge, which is usually close to the bankfull one. Therefore, the channel pattern of each branch must be estimated separately using its bankfull channel width, depth, velocity and resistance. These 'two-echelon' river channels, where a smaller channel is embedded into a larger one, are common at the lowlands of Western Siberia, in the monsoon-influences areas of India, and in the tropical regions of Africa with high variability of flood levels on the rivers.

For the rivers with *TM* diagram of types C and D with one channel-forming discharge the bankfull channel width, depth, velocity and resistance are sufficient for channel pattern estimations.

The climate and landscape determine also floodplain vegetation types and successions. For the river channel pattern, type and growth rate of the pioneer vegetation are of primary importance. During the low water stages bar surfaces are not submerged. In favourable conditions the pioneer vegetation settles on their surfaces. This process takes more than one hundred years in the high Arctic. There are sand bars in the deltas of the Yenisei River and of the Yana River, described at the beginning of the 20th century, which still remain at the same places and state. In temperate climate (for example, on the southern Russian Plain) willow occupies mature sand bars in a few years, and after 10-20 years they become completely stabilised by vegetation and transform into channel meanders or islands, depending on channel /bar width ratio. A well-known example of the Brazos River valley occupation by tamarisk, which caused transformation of the wide braided channel into a narrow sinuous stream, was described by Blackburn et al. (1982).

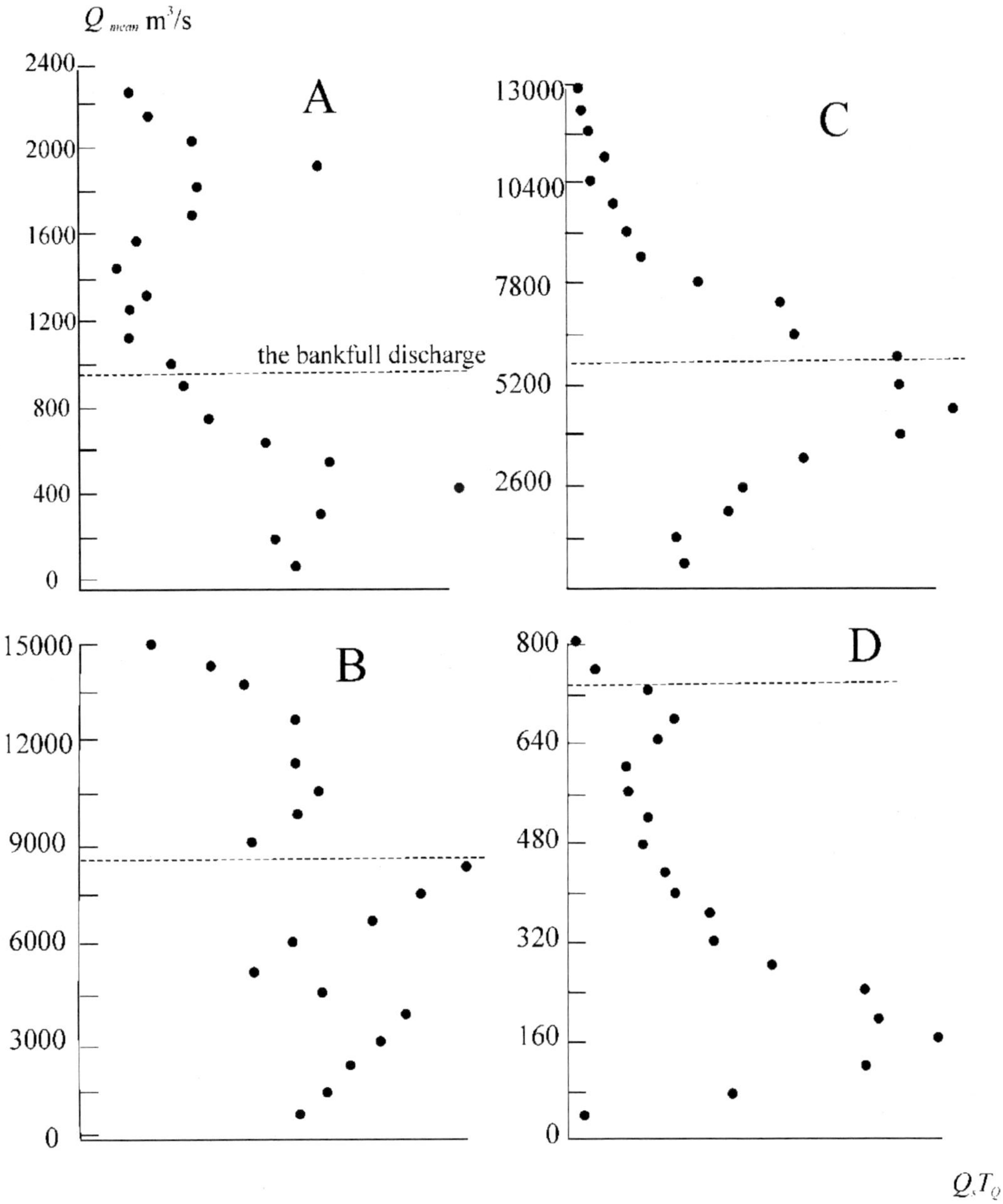

Figure 3.3. The main types of *Q-TM* diagrams for channel-forming discharge estimation (after Chalov, 1979, simplified).

These alluvial features stabilised by vegetated often determine the river channel pattern (channel shape in plan). A river channel is called meandering when the alternating or point bars with the lateral wavelength close to the double width of the channel are vegetated, covered by floodplain alluvium and thus stabilised. Meanders are usually simple when the largest bars of hierarchical complex are vegetated and stabilised. Meanders can be of complicated shape when alternating or point bars with similar relative wavelengths, stabilised with vegetation and fine floodplain alluvium, are not the largest features. In this case, transformation of the undulations with the larger wavelength potentially can lead to compound meander formation. During the low water stages smaller non-vegetated bars emerge, and the river channel can be defined as meandering with alternating and braid bars.

A river channel is called braided by islands when lateral wavelengths of the largest vegetated alluvial features, which determine the channel shape, are less than the channel double width. The larger the channel width is and the smaller the vegetated alluvial features are, the more complicated the braiding is. During the low water stages, the channel can be braided by islands and non-vegetated bars.

Alluvial features under stabilising vegetation cover and fine floodplain alluvium determine river channel patterns most effectively at the rivers with types C and D of hydrological regimen, when flow velocities during the floods are lower than the critical velocity of erosion initiation of the floodplain. At the rivers with types A and B of hydrological regimen channel pattern can be transformed during a high flood, or new channels of the same morphological type can appear due to the floodplain surface erosion.

Geological structure and texture of the river valley often determine a general valley configuration, and even the local geometry of channel bends and braids. Textural (lithologic) composition of the valley often puts a limit on the width of the valley bottom, and therefore on the size of the largest alluvial features. The lower Lena River shows a typical example of such interaction. After the main tributaries – the Aldan River and the Viluy River, flow into the Lena River, the valley bottom of the latter reaches the width of 28 km, its main part being occupied by the channel braided with large islands and bars. At the lower reaches, the Lena River crosses the Verkhoyansk Ridge in a straight antecedent valley only 2 km wide, and only bed forms are developed in this narrow and deep channel.

Large scale spatial heterogeneity of the structural and textural composition of the valley banks also put a limit on the size of the largest alluvial features, as channel bends due to the geological structures can camouflage such alluvial features as meanders.

The evolution of the river valley is imprinted in the valley and channel geometry and deposits. The past channel relief and deposits to a great extent determine the recent processes. The width of the valley bottom is usually the result of the former long-term lateral migration of the channel, the geological structure and texture only temporary limiting its migration. Floodplain morphology is a combination of stabilized alluvial features of various geometry, sizes and ages. Some of these ancient floodplain segments control the geometry of the recent river channel.

The river channel pattern of the Vychegda River (a tributary of the Northern Dvina River) shows a good example of complicated interaction between alluvial relief, vegetation on the sand shallows and geological structure of the river valley (Figure 3.4). At the headwaters, the river is characterised by meandering channel with small submerged linguoid bars of the second order and large alternating bars of the first order. The alternating bars are 2-2.5 times shorter than channel meanders, and linguoid bars are 2.5-4 times shorter, than alternating ones. The wavelengths of meanders, alternating bars and linguoid bars increase down the river with an increase in the channel depth, width and discharge, but their relative proportions remain approximately the same. When alternating bars moving downstream become high enough, their tops become colonised by pioneer vegetation during the low water. These small islands become the cores of stabilisation of the bars during the floods. All well-developed alternating bars become completely stabilised by grass, willows and birch trees. As the result, stabilised high alternating bars with floodplain deposits on the surface transform into channel meanders. In the same time, linguoid bars of the second order become asymmetrical and transform into alternating bars of the first order. Smaller alluvial features take place of the second order bars, so that all the hierarchical system shifts one order up. The length of channel meanders is limited by the river valley geological structure. The bends of the Vychegda River valley bottom are determined by the geometry of depositional bodies of different age and composition. These bends have some characteristic length. When the river meanders grow to the size of these depositional bodies, the hydraulic sinuosity is camouflaged by the structural sinuosity. Smaller alluvial features take place of channel meanders.

Large and the largest alluvial rivers are characterised by more complicated channel relief, firstly, because they are chiefly lowland rivers with low roughness and Froude number. Therefore, they have a potential to form the whole hierarchy of alluvial features, which corresponds to hydraulic conditions. Secondly, they usually flow in broad floodplains, where a channel is not confined by the valley sides. Therefore morphological and geological conditions are favourable for the largest alluvial features formation. Thirdly, flood flow in such rivers is usually powerful enough to submerge and mobilise even large alluvial features. Only the largest alluvial features remain permanently covered and stabilised with vegetation and floodplain silt. Therefore, hydrological conditions are also favourable for large number of alluvial features to be active.

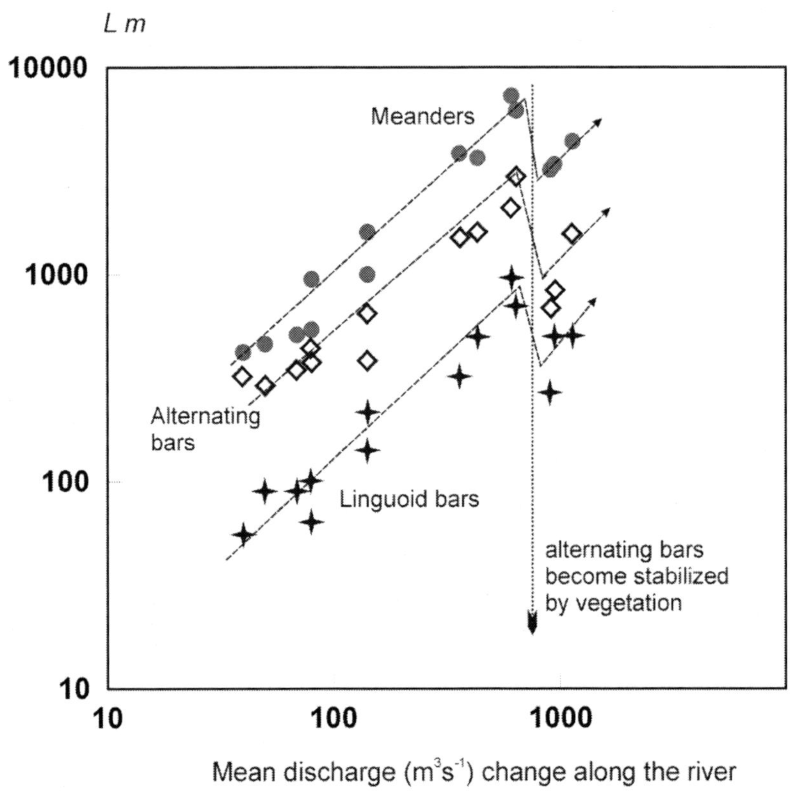

Figure 3.4. Changes of the hierarchical complex of alluvial features along the Vychegda River.

The channel patterns of the small-sized lowland rivers are mostly determined by non hydraulic factors: climate, landscape, river valley geology and evolution. These rivers are characterized by poorly structured channel geometry, which is highly influenced by the local conditions.

The problem of determination of a river channel pattern is not solved theoretically nowadays. Only semi-theoretical or, better to say, semi-empirical approach to this problem is realized. To solve the problem at the semi-empirical level, flow and channel characteristics in the conditions of channel-forming discharge must be available: flow velocity, resistance to flow, channel depth, and channel or valley bottom width. The first three, organized into complexes DFr and λDFr^{-1} determine the hierarchy of alluvial features on the semi-theoretical diagram in Figure 2.25. Channel width or valley bottom width determines the size of the largest alluvial feature, which can be formed in a given channel. Only empirical estimation of channel width is possible now (see Figure 2.22).

The lengths of alluvial features, which can be derived from the diagram in Figure 2.25, are semi-empirical and therefore approximate. They represent the mean (or modal) values for each level in the hierarchy, which, together with empirical variation coefficients, determine the probability distribution function for these lengths. The mean or modal height of alluvial features can be estimated from the empirical diagram in Figure 2.24. These values, together with empirical variation coefficients, determine the PDF for the heights of alluvial features for each level of the hierarchy.

CONCLUSION

According to Velikanov (1954), the variability of the river channel morphology is restricted because the river channel pattern is determined by hydraulic characteristics of the flow. Theoretical analysis of 2D equations of hydraulics with the method of initial instability shows the type of this determination. Channel relief structure is formed by unstable bi-sinusoidal alluvial features with a broad range of longitudinal and lateral wavelengths, which form the discrete-continual spectrum. The surface of the larger alluvial features is complicated by smaller alluvial features, and in the entire relief structure is hierarchical. Five main hierarchical levels of alluvial features can be distinguished by theoretical analysis:

1) level of ultramicroforms, which are the smallest 2D and 3D alluvial features with the longitudinal wavelength increasing with the increase of flow depth and Froude number;
2) level of microforms, 3D isometric small alluvial features, also increasing with the increase of flow depth and Froude number;
3) level of mesoforms, 3D elongated middle-size alluvial features; their longitudinal wavelength increases with increasing flow depth and decreasing channel roughness and Froude number;
4) level of macroforms, 3D elongated large alluvial features with the maximum rate of amplitude growth; their longitudinal wavelength also increase with increasing flow

depth and decreasing channel roughness and Froude number; their elongation is controlled by the intensity of the helical circulation and by channel resistance;
5) level of megaforms, which form long strips along the channel, dividing it into several poorly linked branches.

This simplified channel morphology, based only on the theory, is already very complicated, being the result of transform of the discrete-continual spectrum of bi-sinusoidal alluvial features with different amplitudes and longitudinal/lateral wavelengths. Any real river channel morphology is even more complicated.

The main limitation of the theoretical analysis is the necessity of significant simplification of the basic equations of hydrodynamics and of linearization of the 2D hydraulic equation. This simplification leads to the losses of some important modes of unstable alluvial features. Morphological analysis of the hierarchical complexes of channel relief in various rivers shows an existence of 7-9 levels of alluvial features, which partly correspond to the theoretical ones, and partly supplement them.

Megaripples in natural rivers and in large flumes, as the smallest bed forms controlled by flow characteristics, represent the theoretical ultramicroforms. Antidunes and step-pool systems also belong to the level of ultramicroforms in the flows with large Froude numbers.

Dunes, as the bed forms with sizes in between of ripples and bars correspond to the level of microforms. At least three hierarchical orders of dunes are clearly defined in the natural river channels during the flood stage, when these bed forms are active. Natural dunes are 3D isometric features during the active stage, as the theory predicts, but they are transformed into 2D asymmetric features during the passive stage (at low water). The larger the dune is, the longer the duration of the passive stage can be. Natural dunes form well shaped local maximums in the empirical spectra of channel bed elevations, although such local maximums are absent in the theoretical spectrum. This shows the limitation of the theoretical results.

Bars are the main alluvial features which determine the river channel pattern. In some shallow rivers bars belong to the theoretical level of microforms, but the majority of bars correspond to the levels of mesoforms and macroforms. As the theory predicts, bars are the elongated 3D features. Their sizes increase with the flow depth, but decreases with channel resistance and Froude number. There are two or three orders of bars in the natural river channels, often well-defined in the empirical spectra of the channel bed elevations. The theory predicts only one maximum of macroform amplitude growth, which means that empirical data again show limitations of the linearized theoretical analysis. The second significant difference between the theoretical results and the observations at the natural rivers concerns the limits of the bars' maximum size. According to the theory, the maximum size of the macroforms is determined by flow depth, Froude number and channel resistance. In the natural rivers, the maximum size of bars is restricted by the channel width and resistance. The channel width represents the second internal scale for the river channel, the first being the water depth. The depth enters all theoretical formulas of channel morphology, and the width appears only in the empirical relationships.

Bars can enter an active stage only in high-energy flows, in the mountainous rivers and in the lowland rivers during high floods. Usually bars remain at the passive stage, with triangular longitudinal profile and steep downward slope. The largest bars are often stabilised by pioneer vegetation (willow, reeds, tamarisk, sea-buckthorn) and covered by floodplain fine alluvium.

The complexity of the river channel relief is determined by flow hydraulics and channel morphology: flow velocity, channel depth, width and resistance to flow. The river channel pattern is determined by the same factors, and, additionally, by hydrologic regimen, river valley geological texture and structure, river basin and river valley vegetation characteristics, and by the past events of river valley evolution. These additional factors are less important for the rivers with high-energy flows.

The channel patterns of the middle-sized and large lowland rivers are determined by interaction of hydraulics and landscape characteristics. Alluvial features stabilised by floodplain vegetation and fine alluvium determine the long-term geometry of the channel. Flow hydraulics and channel width determine the size of the largest elements of the channel configuration, i.e. the size of the largest islands or meanders. The size of the smallest geometrical elements of the channel, i.e. the smallest islands or meanders, is determined by the hydrological regimen and the rate of floodplain growth due to fine alluvium deposition and colonization by pioneer vegetation. The more alluvial features of various hierarchical levels are colonized by floodplain vegetation the more complicated the river channel pattern becomes.

The channel patterns of the small-sized lowland rivers with low-energy flows are mostly determined by the climatic, landscape, geological factors and valley evolution. Therefore, these rivers are characterized by poorly structured channel geometry and highly influenced by the accidental combinations of local factors.

REFERENCES

Allen, J. R. L. Polymodal dune assemblages: an interpretation in term of dune creation – destruction in periodic flows. *Sedimentary Geol.* 1978, 20, 17-28.

Berkovich, K. M., Ruleva, S. N., & Chalov, R. S. The channel behaviour of the upper Ob' River. *Geography and Natural Resources* 1989, 4, 54-61 (in Russian)

Blackburn, W. H., Knight, R.W., & Schuster, J. L. Saltcedar influence on sedimentation in the Brazos River. *J. Soil and Water Conserv.,* 1982, 37, 298-301.

Bridge, J. *Rivers and Floodplains*; Willey: Blackwell, 2003, 491 p.

Callander, R. A. 1969. Instability and river channels. *J. Fluid Mech.* 1969, 36, 465-480.

Carling, P.A., & Orr, H. G. Morphology of riffle-pool sequences in the river Severn, England. *Earth Surf. Proc. Landforms* 2000, 25, 369-384.

Chalov, R. S. *Geographic Investigations of Fluvial Processes*; Moscow University Press: Moscow, 1979; 168 p. (in Russian)

Coleman, S. E, & Melville, B. W. Initiation of bed forms in flat sand bed. *J. Hydraul. Engineering* 1996, 122, 64-76.

Fredsoe, J. Meandering and braiding of rivers. *J. Fluid Mech.* 1978, 84, 609-624.

Gregory, K. J. *River Channel Changes;* Wiley: Chichester, 1977, 448 p.

Guy, N. P., Simons, D. B., & Richardson, E. V. Summary of alluvial channel data from flume experiments. *US Geol. Surv. Prof. Pap.* 1967, 462-I, 1-96.

Jain, S. C, & Kennedy, J. F. The spectral evolution of sedimentary bed forms. *J. Fluid Mech.,* 1974, 63, 301-314.

Kartvelishvily, N. A. *The Flows in Non-erodible Channels;* Gidrometeoizdat: Leningrad, 1973, 279 p. (in Russian)

Kennedy, J. F. The mechanics of dunes and antidunes in erodible-bed channels. *J. Fluid Mech.* 1963, 16, 521-544.

Knighton, D. *Fluvial Forms and Processes, a New Perspective*; Arnold: London, 1998, 383 p.

Langbein, W. B., & Leopold, L. B. River meanders – theory of minimum variance. *U. S. Geol. Surv. Prof. Pap.* 1966, 422-H, 1-15.

Lenzi, M. A. Step-pool evolution in the Rio Cordon, northeastern Italy. *Earth Surf. Proc. Landforms* 2001, 26, 991-1008.

Leopold, L. B. *A View of the River;* Harvard College Press: Harvard, 1994, 298 p.

Leopold, L. B., & Wolman, M. G. River Channel Patterns: Braided, Meandering and Straight. *U. S. Geol. Surv. Prof. Pap* 1957, 282-B, 1-51.

Leopold, L. B., & Wolman, M. G. River meanders. *Bull. Geol. Soc. Amer.* 1960, 71, 769-793.

Losievsky, A. I. Laboratory study of riffle formation process. *Trudy MOTSNIIVT* 1934, 86, 1-156 (in Russian)

Makkaveev, N. I. *The River Channel and Erosion in its Basin;* Akademizdat: Moscow, 1955, 346 p. (in Russian)

Mikhinov, A. Ye. Instability of the bottom forms in alluvial channel. *Meteorology and Hydrology* 1983, 11, 84-91 (in Russian)

Nezu, I., & Nakagawa, H. Self-forming mechanism of longitudinal sand ridges and troughs in fluvial open-channel flows. *Hydraul. Environ. 23rd Congr.* Ottawa 1989, Vol. B, 65-72.

Nomenclature for bed forms in alluvial channels. *J. Hydraulic Division, Proc. ASCE,* 1966, 92, 51-64.

Reineck, H.-E., & Singh, I. B. *Depositional Sedimentary Environments, with Reference to Terrigenous Clastics;* Springer Verlag: Berlin, 1980, 549 p.

Richards, K. S. The formation of ripples and dunes on an erodible bed. *J. Fluid Mech.* 1980, 99, 597-618.

Richards, K.S. *Rivers: Form and Process in Alluvial Channels;* Blackburn Press: Blackburn, 2004, 376 p.

Rossinskiy, K. I., & Kuz'min, I. A. Sediment budget approach to calculations of the channel bed deformations. *Trudy Gidroproekta*, 1964, 12, 265-271 (in Russian)

Rouse, H. Experiments on the mechanics of the sediment suspension. *Proc. 5th Int. Congr. Appl. Mech.* Cambridge, MA., 1938, 47-53.

Rozovskiy, I. L. *The water motion on the alluvial channel bend;* AN UkrSSR: Kiev, 1957, 188 p. (in Russian)

Schumm, S. A. *River Variability and Complexity*; Cambridge University Press: Cambridge, 2005, 220 p.

Schumm, S. A. *The Fluvial System;* Willey: Chichester, 1977, 338 p.

Sidorchuk, A. The structure of river bed relief. In*: Coherent Flow Structures in Open Channels*; Ashworth, P., Bennett, S., Best, J. L., & McLelland, S.; Eds.; Wiley: Chichester, 1996; pp 397-421.

Sidorchuk, A. Y. *The Structure of the River Channel Relief*; Gidrometeoizdat: Sankt-Petersburg, 1992, 125 p. (in Russian)

Sidorchuk, A., Schmidt, J., & Cooper, G. Variability of shallow overland flow velocity and soil aggregate transport observed with digital videography. *Hydrol. Proc.* 2008, 22, 4035-4048.

Sinnock, S., & Rao, A. R. A heuristic method for measurement and characterization of river meander wave length. *Water Resour. Res.* 1984, 20, 1443-1452.

van Rijn, L. Sediment transport, part I: bed load transport. *J. Hydraul. Eng.* 1985, 110, 1431-1456.

Velikanov, M. A. *Dynamics of the Channel Flows*; Gostekhiyzdat: Moscow, 1954; Vol. 1, 323 p. (in Russian)

Wang, W. C., & Shen, H.W. Statistical properties of alluvial bed forms. *Proc. 3rd Int. Symp. Stochastic Hydraul.* Tokyo, 1980, 371-389.

Whittaker, J. G, & Jaeggi, M. N. R. Origin of step-pool systems in mountain streams. *J. Hydraulics Division ASCE,* 1982, 108, 758-773.

Xu, J. Comparison of hydraulic geometry between sand- and gravel-bed rivers in relation to channel pattern discrimination. *Earth Surf. Process. Landforms* 2002, 29, 645-657.

Yalin, M.S., & Karahan, E. Steepness of sedimentary dunes. *J. Hydraul. Div. Proc. Amer. Soc. Civ. Eng.* 1979., 105, 381-392.

In: River Channels: Types, Dynamics and Changes
Editor: Daniel A. Molina, pp. 111-136
ISBN 978-1-61324-148-6

Chapter 3

FLUVIAL CLASSIFICATIONS: CONTRIBUTIONS FROM RUSSIA

***Roman S. Chalov*[*]*, Sergey R. Chalov*[†] *and Aleksandra S. Chalova*[‡]**

Faculty of Geography, Lomonosov Moscow State University, Moscow, Russia

ABSTRACT

Though various classifications have been proposed to impose some rational sort of classification on varieties of fluvial processes, only a few could be regarded as universal ones. The most elaborate classification schemes so far derived are from D. Rosgen [Rosgen, 1994], D. Montgomery [Montgomery, Buffington, 1997], and S. Schumm [Schumm 1985]. Besides, system of stream classification and inventory which propose a unique system for linking variety of in-channel processes, forms, and physical and ecological processes, and biological patterns in rivers and floodplains were elaborated at Lomonosov MSU, Russia. Its basic principles (theory of unified fluvial erosion-accumulation process) were firstly formulated by N. I. Makkaveev [Makkaveev, 1955]. The list of classification systems, developed by significant contributors before the early 2000s for both stream and valley types is not small; the given ones represent a new tool in hydromorphological research. It was derived using data from large numbers of rivers from various environments of the former USSR and thus applied both to high-gradient mountain, debris-flow, rocky, plain, and marshy rivers. The classification is obtained by descriptions of the channel forms according to their spatial scale. Fluvial processes and channel patterns types are used to describe channel level of the fluvial classification. Five main fluvial process types in the stream corridor (from high-gradient streams with rapid/falls to plain rivers) and 3 groups of channel patterns (straight, meandering, and braided) are found. All of the types are separated using a variety of quantitative geomorphologic indicators (stream order, stream power, stream size and channel slopes, entrenchment ratio, and sinuosity or amount of branches).

[*] rschalov@mail.ru
[†] srchalov@geogr.msu.ru
[‡] aleksandra-1984@mail.ru

The above-mentioned classifications are much different in the given channel type's quantitative and qualitative descriptions. Variety of views on channel types determines that uncertainties exist in fluvial system management between different researchers and practitioners. The underlying problems is to find the unique approach to the variety of channel forms, to derive the complete analytical and empirical solution for the sediment entrainment, transport and deposition and to develop the specific assessment of stream stability, channel evolution, and potential response for each channel type. The goal of this paper is to estimate the state-of-the-art methods in fluvial classification and describe the mostly full approach to detail the variety of rivers and channels. Our final conclusion that river channels should be described through comprehensive qualitative and quantitative descriptions of the all its controls, forms, and constituents. The complex approach enables researchers to find places for the object and thus classify it or put it in an order and further apply this information for channel evolution, fluvial management, and safety maintenance. The corresponding examples are given in the paper.

INTRODUCTION

Fluvial processes describe the interactions of fluid water and erodible materials in the channel boundary through the entrainment, transport, and deposition of sediment and determine the temporal changes of the morphology of natural rivers. Together with catchment processes, fluvial processes are the main agent of global denudation and matter movement and, finally, play a key role in environment. Vulnerability of the rivers issues a challenge to study fluvial processes. Anthropogenic uses along stream corridors increase greatly, as do channel disturbances. Cumulatively, disturbances may alter channel physics, impacting biological structure and diversity, and inducing instability. The negative effects of human-induced practices have created a need for understanding, assessing, and restoring natural types of fluvial processes. At the same time human activities strongly depend on water presence; the latter usually influences safety of buildings, constructions, dams etc. Channel erosion, avalanches, channel cut-offs, channel aggradation, or degradation prevents use of water resources. Necessity to manage and protect fluvial systems determines growing demand to study fluvial processes.

Rivers remove water and sediments supplied to them, and through this produce a great variety of channels. In fluvial geomorphology, rivers of this variety are the key point for channel evolution prediction, paleogeographic studies, water management, fish habitat control, etc. Classifying natural streams still challenge researchers as needs and knowledge grow. Though various fluvial classifications have been proposed to impose some rational sort of methodology for rivers and channels, none of them have been regarded as the universal one. The most elaborate classification schemes yet derived by are D. Rosgen [Rosgen 1994], D. Montgomery [Montgomery, Buffington, 1997], and S. Schumm [Schumm, 1985]. Several recent reviews of fluvial classification systems have been published [Naiman et al., 1992; Downs, 1995; Kaszowski, Krzemien, 1999], but none of the English language works deal with the results of the abundant studies of the fluvial processes in Russia. Besides, none of them cover all the types of the environments that could be found in the vast territory of Russia. The detailed classification of the fluvial processes based on analysis of various rivers of the USSR was derived in Moscow State University on the basis of Nikolay I. Makkaveev's [Makkaveev, 1955] theory of unified fluvial erosion-accumulation processes. It was in demand at many stream engineering projects in the former USSR and Russia [Roman S. Chalov, 1979, 2008], and thus was elaborated strongly. Makkaveev-Chalov classification is based on the observations in different environments of Russia – from mountains (Caucasus, Than-Shan etc) to flatlands (East European plane, West Siberian plain, etc), and both at ephemeral and perennial nival belts.

The above-mentioned classifications are similar covering both mountains and plain rivers, but are much different in the given channel types quantitative and qualitative descriptions. Variety of views on channel types determines that uncertainties exist in fluvial system management between different researchers and practitioners. The underlying problem is to find the unique approach to the variety of channel forms, to derive the complete analytical and empirical solution for the sediment entrainment, transport and deposition, to develop the specific assessment of stream stability, channel evolution, and potential responses for each channel type. The goal of this paper is to introduce Russian classification to the English-spoken World and to estimate the state-of-the-art methods, including the contribution of Russian sciences and their comparisons . The special concern of this paper is to describe the application of Makkaveev-Chalov classification. For brevity, we evaluated only a few better known English-written fluvial classification schemes and the corresponding approaches together with detailed explanations of the MSU classification. The distinctions come from estimating control variables, their role, and linking with channel forms variety. Thus, the control variables of primary importance further lead to correct describing and classifying of the object and channel evolution predictions.

Table. 1. Total scheme of the channel evolution factors

<table>
<tr><td>Type</td><td colspan="2">Factor</td><td colspan="2">Characteristics</td></tr>
<tr><td rowspan="2">Inputs</td><td colspan="2">Water runoff</td><td colspan="2">Water discharge, water regime, effective discharge</td></tr>
<tr><td colspan="2">Sediment transport</td><td colspan="2">Bed load/suspended load ratio, sediment load</td></tr>
<tr><td rowspan="9">Constraints and impacts</td><td colspan="2">Valley geology</td><td>Deposits composition and density</td><td rowspan="3">Channel stability, longitudinal profile</td></tr>
<tr><td colspan="2">Valley geomorphology</td><td>Slope, valley type, valley bottom and channel width, entrenchment ratio</td></tr>
<tr><td colspan="2">Bed and floodplain (valley) material</td><td>Particle size</td></tr>
<tr><td rowspan="6">Local</td><td>Ice events</td><td colspan="2">Frequency and type</td></tr>
<tr><td>Riparian vegetation</td><td colspan="2">Density and type</td></tr>
<tr><td>Woody debris</td><td colspan="2">Amount and frequency of logjams</td></tr>
<tr><td>Stream biota</td><td colspan="2">Amount and density of fish assemblages, spawning-driven sediment transport</td></tr>
<tr><td>Permafrost</td><td colspan="2">Deposits freezing, distribution of subglacial taliks</td></tr>
<tr><td>Wind</td><td colspan="2">Amount of windblown sediments, wave height</td></tr>
</table>

Controls of Channel Processes for Classifying Natural Rivers

River channels are controlled by interactions among a set of constant variables. The main task for recognizing all kinds of natural rivers is to identify the significant variables and determine how they control river channels. Many fluvial classification schemes have been based upon characteristics of channel form. Channel forms, although readily measurable, are the end products of the complex. The better way is to study interaction of channel forms and factors of channel origin and evolution. This approach is partly implemented in all concerned approaches.

The list of the features to identify stream types in D. Rosgen's classification (Rosgen, 1994) system are : 1) single or multiple threaded channels (three or more channels at bankfull discharge); 2) entrenchment ratio; 3) width to depth ratio; 4) sinuosity; 5) slope ranges, and 6) channel material. As one can see, there are 3 control variables (3, 5, 6) and 3 geometric features (1, 2, 4). The main disadvantage of the scheme is the lack of some variables which are regarded to be the inputs of the fluvial system. Improvement of this scheme could be done through the increase in the amount of original factors. It is well-known that channel evolution is determined by the stream size (water runoff), sediment load (suspended and bed load), valley confinement (geologic and geomorphologic factors), slope, bed material, and many local factors – ice events, riparian vegetation, woody debris, and stream communities. They could be divided on the Inputs and Constraints/impacts (table 1). The following links between controlling variables cover the broadest range of factors influencing the fluvial processes: 1) stream size and channel slopes determine type of river (from mountain to plain rivers); 2) entrenchment ratio; 3) channel pattern type determined through the different sinuosity, amount of branches and channel shape; 4) bed forms (list of features is specific for straight, meandering and braided channels), and 5) channel material and sediment load.

Input

It is evident that the dominant controls of channel forms are water discharge and sediment load. Being the primary control parameter, water discharge has not proved to be an effective criterion for classification (Knighton, 1998). Nevertheless, jointly with the main constraint factor (channel slope), water discharge continues to be the universal parameter for identifying channel types through gradient–discharge chart analysis because discharge (*Q)* and slope (*s)* are the two variable quantities in stream power. The physical essence of this chart is that the higher and further to the right a river plot is, the higher stream power it has, because a high value of both discharge and slope cause a high stream power. Practically all research of this kind has shown that meandering channels plot lower than multiple ones [Alabyan, Chalov, 1998].

Wide ranges of relationships to divide channel pattern types were constructed using not only hydraulic factors, but also sedimentary ones [Knighton, 1998; Chalov, 2008]. Romashin [Romashin 1968], Carson [Carson 1984], and Ferguson (Ferguson, 1984] were right to contend that the valley (but not the channel) gradient and the magnitude–frequency definition of discharge must be used to secure independence of the control variables. Originally, the use of magnitude–frequency analysis to calculate a representative discharge was suggested by Schaffernak [Schaffernak 1950], who investigated the water level spectrum and the product of stream power and frequency. Critical reviews of such investigations were contained both in Russian (e.g. [Makkaveev, Chalov, 1986]) and English-written papers (e.g. [Carson, 1984]).

Also, there were a lot of discussions about what discharge is more representative and should be taken for such analysis. Rosgen's stream classification system [Rosgen 1994] is based on geomorphic measurements and indicators taken at bankfull discharge. It aroused a lot criticism because a returns interval of 1.5 years was used to determine bankfull variables in Rosgen's system without success. Despite this, further investigations [Southerland, 2003] found that bankfull dimensions are easily identifiable (e.g. on both the east and west slopes of the North Cascades in Washington State).

N.I. Makkaveev [Makkaveev, 1955] and L. Wolman and Miller [Wolman, Miller, 1960] postulated that in alluvial rivers, the discharge which transports the greatest sediment in the long term is the effective (also so-called "dominant" or "channel-forming") discharge. Practically, the effective discharge may be determined using a diagram in which values of the product of sediment discharge and frequency of occurrence are plotted against corresponding water discharge values – *the effective discharge curve*. The effective discharge has the maximum value of this product and corresponds to the maximum of the curve. According to Makkaveev, rivers of the Russian Plain usually have one or several effective values, one of which is close to the mean annual flood. This fact confirms the possibility of using the mean (or median) annual flood as the first approximation of the effective discharge when constructing discharge–slope (Q–s) diagrams.

The above mentioned effective discharge curve (EDC) is an integral characteristic of hydrological and sediment regimes, and reflects the main features of channel patterns. The EDC may be used for channel pattern analysis. Every flow that transports sediments affects the channel form. Product of sediment discharge and its frequency may be considered an index of channel forming rates. Constructing the EDC curve allows conversion from solitary discharge values (mean, extreme, bankfull etc.) to the continuous discharge spectrum.

To compare EDC features with typically used Q values, several horizontal lines may be drawn including some which correspond to certain transitions [Alabyan, Chalov, 1998]:

1. 'the bankfull line' corresponds to the bankfull discharge;
2. 'the meandering–branching threshold line' corresponds to Qs=0.4 (s value is supposed to be a constant);
3. 'the sidebar–mid-bar threshold line' corresponds to Qs=0.1.

Correlation of EDC elements with these three lines indicates a relationship with channel features, and may be illustrated by the following examples (Figure 1). All of these rivers are located in the West-Siberian Lowland, composed of soft sedimentary rocks allowing free conditions for channel development.

The Upper Ob River has a complicated split channel, with medial bars prevailing in branches. Almost the entire EDC lies below the bankfull line (Figure 1A), so channel-forming activity is concentrated below the floodplain level. Both the high-water and low-water part of the EDC are located above the sidebar–mid-bar threshold line, so mid-channel bars may be formed during every hydrological season. The high-water part of the EDC is situated above the meandering–branching threshold line, so stream power during a flood is sufficient to form splits in the channel.

The Ob River in its lower reach is characterized by anabranching (branches are meandering with alternating bars). The upper maximum of the EDC is well encountered and located above the meandering–branching threshold line which in turn plots above the bankfull line (Figure 1B). So, the power is sufficient to cause channel division only when the floodplain is inundated. The Chulym River (a tributary of the Middle Ob) has a meandering channel with alternating side bars and individual mid-bars. The meandering–branching threshold line is situated above the EDC (Figure 1C) so the channel is single-thread. Mid-bars may be formed only during a flood, but are not maintained through the low-water season. More evidence of empirical approach proves that the stream size determines channel pattern through the absence of the braided channel among the smallest plain rivers[Smirnova, 2002].

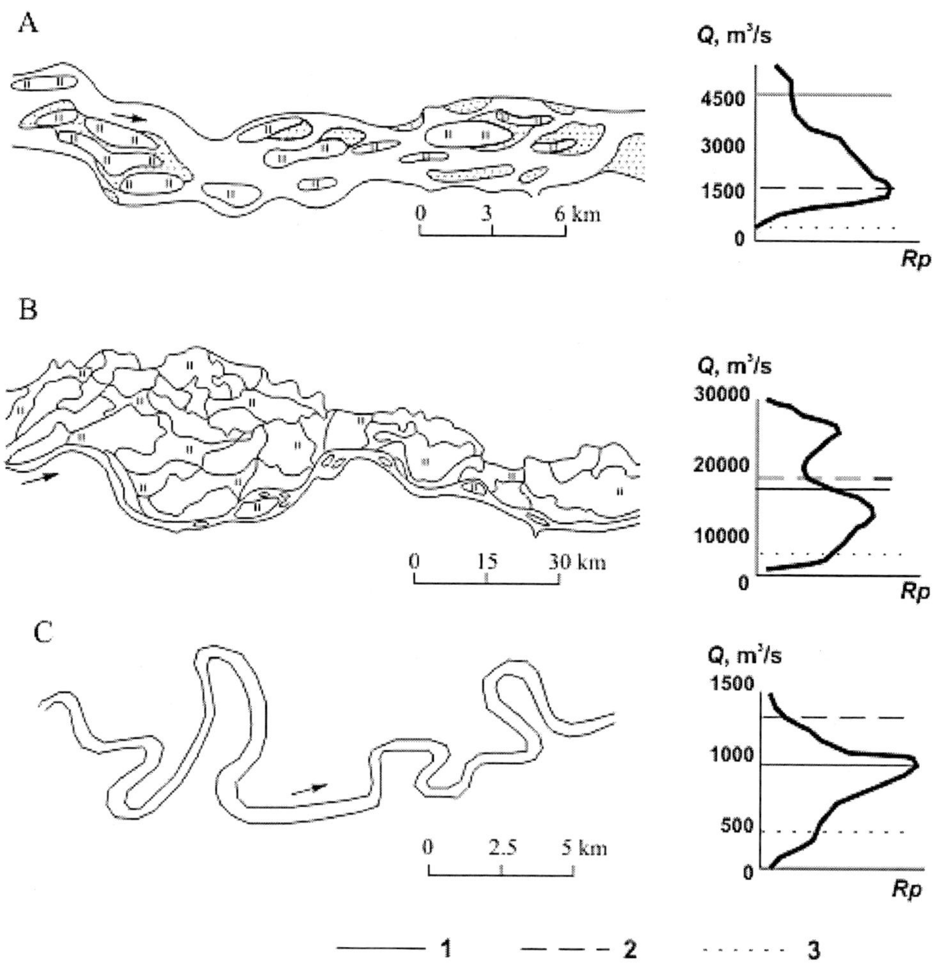

Figure 1. Channel patterns and corresponding effective discharge curves: A – island-braided channel pattern, B – anastomosing channel pattern, C – meandering channel pattern; 1 – the bankfull line; 2 – the meandering–branching transition line; 3 – the sidebar–mid-bar transition line [Alabyan, Chalov, 1998].

Magnitude-frequency analyses were reflected in Makkaveev-Chalov's classification where different stream sizes and slopes are ranged according to channel types (so called "fluvial processes types"). Montgomery classification is based on the observations on small rivers and is applicable only for mountain drainage basins. Absence of the stream size in any of classifications could lead to uncertainties connected with identifying streams from different environments as the same type.

The type and amount of sediment loads moved through the channel is regarded as a basic criterion in Schumm classification [Schumm. 1977, 1985]. The criterion is based on the descriptive base. Bed load rivers are characterized as highly active and low sinuosity bearing channels that exhibit a braided channel pattern. Suspended sediment load dominated rivers are characterized by a single channel with a higher sinuosity, and stable banks because of cohesive fine-grained deposits. Mixed-load rivers exhibit intermediate properties. Evident relationships between sediment transport and channel form were stated by Alekseevsky [Alekseevsky et. al., 2009], who observed the correlation between channel patterns and sediment yields of different rivers. The small and middle rivers usually have meandering channels and specific features (see below), which are closely connected with the bed load/suspended load ratio. Largest rivers develop braided patterns caused by the increase of bed load. The complexity of braided patterns increase with the bed load/suspended load ratio growth. This is apparent from comparison of braided reaches on the Yangtze River (5-6 % at fine sand material prevalence) and Lena River (40 % at coarse sand prevalence).

The other way to involve sediment factors for stream classifying is sediment balance estimation. Channel shape strongly depends on the sediment budget. Meandering channels appear to be the most stable pattern, as the resulting value of sediment budget is close to 0 ($\Delta W \approx 0$) along the entire length of channel sections. For relatively straight single-thread channel pattern conservation, a few conditions should be satisfied. One of those is the excess of the flow transporting capacity over transit sediment flux ($\Delta W > 0$). The opposite situation of the deficient flow transporting capacity ($\Delta W < 0$) is generally typical for multi-thread channel pattern. Exact recognized channel types correlate with sediment balance to a certain degree. The correlation's significance reflects various classification correspondence to sedimentary factors.

CONSTRAINTS

Geology and geomorphology of the river valley strongly effects morphology of the valley and river channel (width, depth, slope) and deposit composition (deposits type and density, bed material size, erodibility). Bed material size is usually assessed through grain size (D_{50}) and types of particles (from silt to boulders and bedrock). It is closely connected with channel stability (bank erodibility). A few equations to assess stream stability in Russia were proposed. N.I. Makkaveev [Makkaveev, 1955] argued that a channel stability index should be used in dimensionless form: $K_s = b/(dI)$, where b stands for channel width, d – bed material mean grain size, and I stands for water surface gradient at mean maximum discharge. K_s values correspond to the length of the largest bed forms (riffles), relative height range, and migration rate.

More evidence for the channel stability is width/depth ratio. This geometric feature increases with the bank erodibility and bed-load transport increases , so it is often used as the control variable for the channel patterns division. D. Rosgen [Rosgen, 1994] and N. Rhzanitcyn [Rhzanitcyn 1985] used b/h ratio to divide different channel pattern types. Braided reaches are characterized by a high width/depth ratio (usually $b/h > 80$), whereas meandering rivers are distinguished by b/h= 50–80 and straight – $b/h < 50$ [Rhzanitcyn 1985]. Evidence of regional data impacts on the fluvial classifications illustrates a distinction from b/h ranges proposed by Rosgen [Rosgen, 1994]. Also, observed misunderstandings could be explained by different approaches to measure channel width to mean depth. Thus, Rosgen used a ratio of bankfull channel width to bankfull mean depth and Rhzanitcyn – a ratio of channel width to maximal channel depth corresponding to the value of effective water discharge. In any case, it proves the uncertainties of morphological data for classifications.

The entrenchment ratio is defined both by Makkaveev-Chalov [Makkaveev, Chalov, 1986] and Rosgen [Rosgen, 1994] as the ratio of the width of the flood-prone area to the bankfull surface width of the channel. The flood-prone area is defined as the width measured at an elevation which is determined as twice the maximum bankfull depth. Entrenchment ratios are usually defined as the flood-prone width divided by the bankfull width. According to Rosgen, entrenchment ratios of 1–1.4 represent entrenched streams, 1.41–2.2 represent moderately entrenched streams, and ratios greater than 2.2 are slightly entrenched (well-developed floodplain). Ratios suggested by Makkaveev-Chalov for low confined channels (with wide floodplain) were >3, for medium confined – 1–3, for highly confined (incised channels) – <1.

None of the local factors could be regarded as criterions. Nevertheless, they often have a strong influence on channel development. The main factors are the vegetation, biota, ice, and wind. There has been increasing interest in the role of vegetation in fluvial geomorphology in recent years [Gregory, 1993] because it has been recognized that river dynamics cannot be fully understood without taking into account the impact of vegetation, both on the banks and within the channel. It impacts bank stability and channel geomorphology, especially of small rivers.

Woody vegetation at small rivers usually becomes the main constraint of channel shifts. It promotes occurrences of waving shapes of the bank line due to the lack of stream capacity to destroy woods. Between the single riparian trees, streams wash the banks, but if the space between the trees is less the channel width, the straight channel pattern occurs. Otherwise, single trees cause flow turn and further bends formations . Meandering channel patterns can be formed under these conditions.

Important sequences of the riparian vegetation is the woody debris forming. The location and storage of woody debris and amount of jams per channel length correlates with the slope, discharge, and length of presence of wooded riparian zones which are the sources of vegetation. Spatially unpredictable debris input processes such as windthrow and beaver activity cause a small percentage of jams. The majority of woody material enters channels through bank erosion, much of which results from mass failure in degraded reaches. Only along low turbulent rivers are beaver's activities becoming the only factor of woody input into rivers. The favorable places of the jams forming correspond to different stream reaches and in general reflect the fluvial processes' continuum along rivers. The unequal possibility to occur at different channel types determines various impacts of jams on stream morphology. It was found (Esin, Chalov, in print) that the largest changes of bed morphology exist on braided rivers of piedmont zone.

The most well-known example of biota influence on river channels is the beaver's activities [Bolisov, 2007]. Recent data shows that spawning of salmon fish (Oncorhynchus) could also play a vital role in fluvial system development. During spawning seasons, salmon spawning leads to a many-fold increase in sediment load due to spawning-generated changes in bed surface topography. This phenomena was described simultaneously for Canadian [Hassan et al, 2008] and Russian rivers [Chalov, 2008]. The latter investigations carried out in 2009 showed that during years when the abundant salmon population is observed, the total sediment load could be increased several times due to mass-spawning salmon. Many-fold increases of both suspended and bed loads occurs. Observations on the Karaga river (East Kamchatka peninsula) showed that the spawning of gorbusha (*Oncorhynchus gorbuscha*) caused up to 40 cm changes of bed load surface on average. Due to the destruction of well armored surfaces, it results in significant bed erosion during floods. The total sediment transport increases abundantly, as do the channel pattern transformations.

Ice jams and ice crusts significantly impact channel patterns. The morphology and landscape of the extensive river valley reaches could be determined by ice events [Makkaveev, 1955]. Special types of alluvial fans occur which should be regarded as special stream types. Little information has still been gathered about wind impacts on fluvial systems. Wind has been observed to mobilize significant quantities of sediment along the river, particularly during summer dry periods. Discharge redistribution in the Upper Ob braided reaches were explained [Makkaveev et al., 1966] by windblown sediment transports from the left to right banks according to prevailing wind direction. The effects of wind in sediments along an island-braided, gravel-bed river were discussed by A.M. Gurnell et.al. [Gurnell et al, 2008].

The given examples illustrate the role of local factors on channels and prove the necessity of fluvial classifications in channel evolution prediction.

Fluvial Classifications

General Description

Stream classification is a necessary planning tool for natural channel restoration and water resource use. Because stream types vary in their bed load transport capabilities , morphology, and their physical relationship with the floodplain, it is necessary to plan accordingly for that potential. Understanding the stream site potential should be a cornerstone to early phases of all human activities (both restoration and changing) on rivers. There are a few main rules for stream classification that were found and stated by many scientists:

1. Stream classifications should be relied on the range of all control variables. A lot of criticism was made on designating one factor as the main control. Thus according to Schumm's theory of sediment load role in channel pattern formations , channel patterns are organized into bed-loads, sediment-loads, and mixed-loads. It is evident

that it does not apply well to all rivers [Van der Berg, 1995]. Finally it should be postulated that sediment control channels form through a variety of secondary events and processes. This proves the necessity of using all factors for stream classification.

2. Valley-channel systems represent a variety of multi-scaled objects that should be divided on different levels of classification . The main task is to divide objects which are differed by spatial location and to make them relevant to the different elevation and climate zones.

 Usually the following components of fluvial systems were classified:

 - channel reach types, based on bed morphology, sediment transport processes, sediment supply, and discharge [Montgomery, Buffington, 1997; Makkaveev, Chalov, 1986];
 - channel patterns (channel form adjustment in horizontal plan) [Leopold, Wolman, 1957; Makkaveev, Chalov, 1986; Schumm, 1985];
 - bed forms types (such as described by Brice and Blodgett [Brice, Blodgett 1978] four channel types of: braided, braided point-bar, wide-bend point-bar, and equi-width point-bar);
 - objects from different structural levels of fluvial relief (such as meandering channel patterns and bar-braided channels in Schumm classification).

 Fluvial classifications as the instrument of the environmental research should express the variety of channel forms and processes. Furthermore, the described objects should be put into clear order by origin, size, and processes. Rosgen [Rosgen, 1994] and Makkaveev-Chalov [Makkaveev, Chalov, 1986] classifications were the only designations in fluvial geomorphology to classify structural levels of fluvial relief in general.

3. Stream classification should classify both processes and forms. Rosgen postulated that form and process are not mutually exclusive as they are critically linked and must be used interchangeably. According to Makkaveev [Makkaveev, 1955], the principles of interchanges between channel pattern and stream flow determines stream evolution. River bend or bifurcation corresponds to the circulating waves which further impact the reach stability, curvature increases, and new islands formation. The Makkaveev-Chalov [Makkaveev, Chalov, 1986] classification was named as "morphodynamic" because it regards both outward signs and the interior stream dynamic.

Comparison

To date, the most widely accepted World classification system used for natural channel restoration is Rosgen's [Rosgen, 1994] Stream Classification System. Stream reaches are divided into 7 types according to exact ranges in sinuosity, entrenchment, width/depth ratio, and gradient. The discussion about the advantages and disadvantages covered the famous Rosgen work and classification. This discussion posed both the above-mentioned necessities for the classifying of streams and confirmed that there are a few blank spots in fluvial classification. Montgomery and Buffington's study [Montgomery, Buffington, 1997] is a good example of a regionally based classification system involving many factors. It was used on the west slopes of the Cascade Mountains in the Pacific Northwest. This system focuses on channel form and process. Stream networks are divided into channel reach types, based on bed morphology, sediment transport processes, sediment supply, and discharge. Montgomery and Buffington's system uses bed morphology or slope as an indicator of channel reach type.

Moscow State University classification is based both on descriptive (qualitative) and on quantitative characteristics of stream size, channel slope, deposits types, etc. It was derived using data from large numbers of rivers from various environments of the former USSR and thus applied both to high-gradient mountains, debris-flow, rocky, and plain marshy rivers. The classification is obtained by description of the following objects according to their spatial scale: stream (fluvial processes) type, entrenchment ratio type, channel pattern type, and bed forms (step-pool or riffle-pool sequence). They form under the given background conditions (topographic megaforms and natural territorial complexes) (Figure 2).

Fluvial processes and channel pattern types are used to describe channel levels of the fluvial classification. Since discharge and slope are the two variable quantities in stream power, the empirical data implies that the changes of river types (so called "type of fluvial processes") can be associated with increasing stream power (table 2). As the criterion for deciding between plain rivers and mountain rivers, Makkaveev suggested the value of critical grade I_{cr}, because its relationship with the actual grade I determines a quiet ($I < I_{cr}$, $Fr < 1$) or violent ($I > I_{cr}$, $Fr > 1$) state of the flow and, hence, the mechanism of its interaction with the channel. In the relations (the parameter of flow kineticity), h is the depth of the flow and is a correction to the velocity.

Channel confinement is characterized through floodplain/channel width. For confinement ratios, we propose to discern incised rivers and rivers with wide floodplains, as well as transitional types of rivers with adapted channels. Rocky channels represent special types of channel confinement. They occur on mountain rivers as well as on plain rivers of different grades. Constriction of the channel by rocky banks and bottoms, and at small grades, causes the channel flow to turn into a rapid stream whose velocities greatly exceed the scouring rates, even for pebble- boulder channels. This ensures the removal of the arriving pebble-boulder (or sand) material beyond the channel.

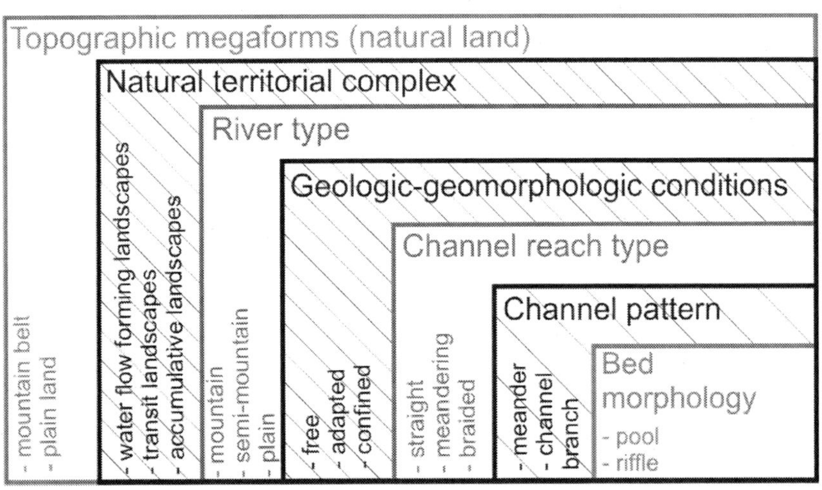

Figure 2. Structural classification of fluvial systems.

Table 2. Type of fluvial processes

Type of fluvial processes	Slope ranges (‰) of rivers with basin area (km^2)			
	>1000	100-1000	10—100	<10
Plain	<0.6	<0.8	<6-7	<14—15
Semi-mountain	0.2-7	0.5 – 7	6-25	15-40
Mountain				
Channel with developed alluvial forms (bars)	1.5-14	5-17	7-30	15-80
Channel without developed alluvial forms (without bars)	3.5-20	7-40	18-70	25-125
Step-pool channel	>10	>20	>25	>40

Finally, each river type represents a certain channel pattern. Different authors consider the original division of the rivers on straight, meandering, and braided to still be relevant, but the further subdivision determines the final content of each classification.

Few parameters have been used to determine certain types of straight, meandering, and braided channels. The crucial task was to divide straight and meandering channels. According to Makkaveev and Chalov's scheme [Makkaveev, Chalov, 1986], the difference can be found through the morphology of straight channels and bends. Statistical analysis showed that even small increases of curvature at slight curves (l/L = 1,15) leads to many-fold increases of cross-section asymmetry ratios defined as $d_{concave}/d_{convex}$, where d is the distance from concave and convex banks correspondingly. That is why the value l/L = 1,15 is considered to be the threshold for the meandering process beginning and transformation of the straight channel to the meandering one.

The group of straight channels includes rather extensive reaches of river channels with relatively straight banks with several gently sloping bends and small islands which do not make stable alterations of the flow structure downstream and do not provoke its changes across channel width. Variability of natural factors determine the existence of different types of straight channels due to river valley morphology – free, confined, and incised channels (Chalov, 2008).

Few main types of meandering channels according to geologic-geomorphologic conditions of meanders development and meanders configuration have been recognized: free, confined, and incised meanders (Figure 3.1); loop-shaped, segment-shaped, sinusoid-shaped, etc. (Figure 3.2) [Makkaveev, Chalov, 1986]. Being originally morphologically induced, these types present a variety of stream hydraulic and stages of evolution. Segment-shapes are distinguished by narrow bend spurs and counter bank erosion which determines trends for the cut-off. Sinusoid-shapes are characterized by straight reaches on the bend limb, whereas "lopsided" bends form due to constraint factor impacts (mostly deposits composition). The specific transitional form between meanders and braided channels are cut-off, and further development depends on the stability of former river-bed.

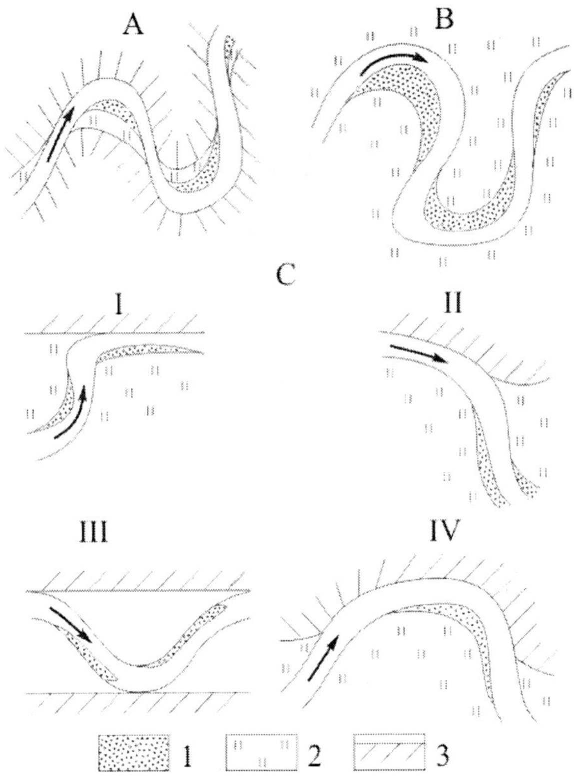

Figure 3.1. Types of meandering channels due to geologic-geomorphologic conditions: A – incised meanders, B – free meanders, C – confined meanders (I –with the upper meander limb near bedrock bank, II – induced meander, III – meander in contact with bedrock bank, IV – insert meander). 1 – side-bars, 2 – flood plain, 3 – bedrock bank.

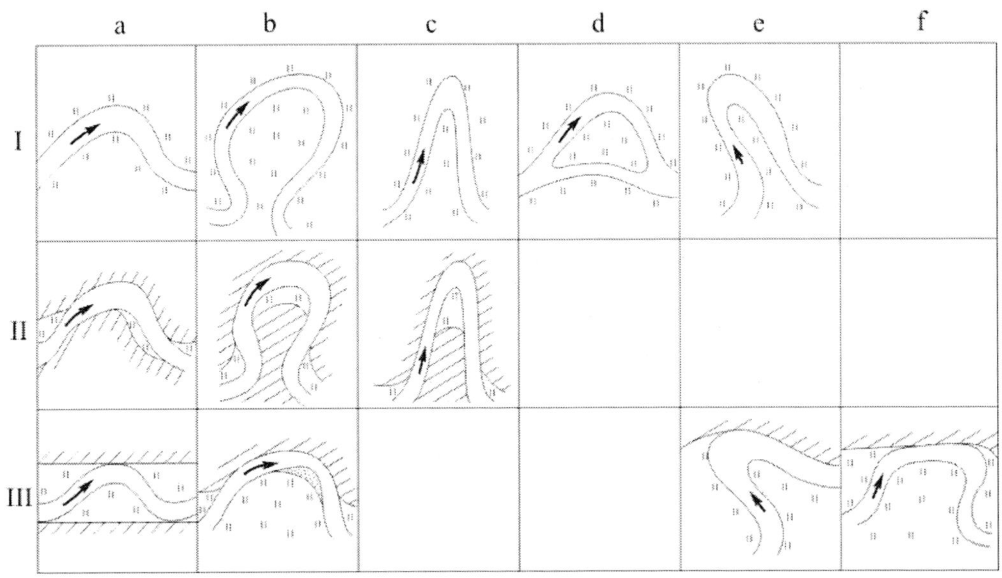

Figure 3.2. Types of meandering channels due to their plan configuration: a – segment-shaped meander, b – loop-shaped meander, c – sinusoid-shaped meander, d – cut-off meander, e – lopsided, f – box-shaped meander. (I – free meanders, II – incised meanders, III – confined meanders).

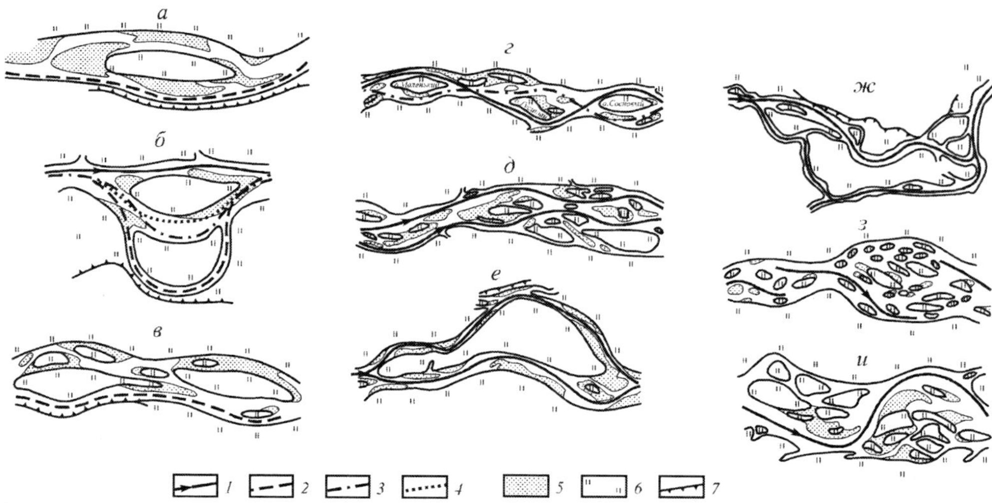

Simple: *а* – single; *в* – one-side. Complex: *б* – complex single; *г* – conjugated; *д* – parallel-braided; *е*, *ж* – anabranching; *з* – complex conjugated; *и* – complex one-side
Dynamic axis of the flow: 1 – modern; 2-4 – previous; 5 – side bars; 6 – floodplain; 7 – valley

Figure 4. Island-braided and anabranching channels [Chalov, 2008].

Island-braided channels represent a variety of forms (Figure 4). The morphological descriptions of the main types are the primary feature of Makkaveev-Chalov classification. The simple and complex reaches are identified. So-called “Single” and “one-side” braided reaches form the first group are also identified. Complex braided channels include 6 types. There are “conjugated” braided reaches where water flow crosses the river between banks (so called “figure of eight rule”). Flow pattern at braided “complex conjugated” and “complex one-side” reaches where flow crosses the channel among large amounts of relatively small islands. All types of conjugated braided reaches are distinguished by discharge redistribution in the case of flow shifts among upper branches. Two or more general branches separated by long islands and connected by small streams form essences of “parallel-braided” reaches.

The shape of braided channels have much in common with anabranching channels. Nevertheless, the latter are distinguished as the upper structural level of fluvial relief. Similarly bar-braided channels are the lower level of fluvial relief. In anabranching rivers, islands usually persist for decades and centuries, and are many times wider than the channels. Thus, each branch of the anabranching system can be characterized by specific channel patterns (straight, meandering, or island- braided). On the basis of various hydraulic and physiographic conditions, Knighton [Knighton, 1998] proposed the most elaborate classification of anabranching channels which recognized six types. The special type (“doubled channels”) was described by Chalov R.S. [Chalov R.S., 2008] to identify extensive braided reaches divided by large islands. The latter corresponds to the greatest braided reaches of the World rivers (length up to 300-500 km as on the Ob river).

We compared the three most elaborate classifications proposed by Rosgen (1994), Montgomery-Buffington [Montgomery, Buffington, 1997], and Makkaveev-Chalov [Makkaveev, Chalov, 1986; Chalov, 2008] (table 3). The results of the comparison showed that even the most elaborate classifications suffer from the conditional conclusions. For the best result, the final definition of each stream type must consist of river-confinement-channel pattern type’s description and detailed identification of the controls variables. River channels should be described through comprehensive qualitative and quantitative descriptions of all its controls, forms, and constituents. Nevertheless, stream types together with cross-section, bank erosion, sediment data, riparian vegetation, and fisheries data can be used to predict the risk and sensitivity to disturbance.

Table 3. The general correspondence between the main stream types of Rosgen, Makkaveev-Chalov, and Montgomery classifications

MSU classification			Montgomery classification						
			Alluvial					Colluvial	Bedrock
Fluvial processes type	Confinement	Channel type	Cascade	Step-pool	Plane-bed	Pool-riffle	Dune-riffle	Colluvial	Bedrock
Mountain	Bedrock	Straight	A1a+, G1	F1					
		Meandering							
		Braided							
Mountain with rapid/falls	Confined	Straight	A2a+, G2	F2					
Mountain without alluvial forms	Confined	Straight			Aa, A				
		Meandering			G, Gc				
		Island-Braided	Db, D						
Mountain with alluvial form	High (floodplain is absent)	Straight	Aa, A						
		Meandering	Gc, G, Fb, F						
		Island-Braided	Db, D, Dc, Da						
	Medium (narrow floodplain)	Straight	Ba, B, Bc						
		Meandering	Ba, B, Bc						
		Island-Braided	Db, D, Dc, Da						
		Anabranching	D, Dc, Da						
	Low (wide floodplain)	Straight				Ca, C, Cc			
		Meandering				Eb,E			
		Island-Braided	Db, D, Dc, Da						
		Anabranching	D, Dc, Da						

Table 3. Continued.

MSU classification			Montgomery classification						
			Alluvial					Colluvial	Bedrock
Fluvial processes type	Confinement	Channel type	Cascade	Step-pool	Plane-bed	Pool-riffle	Dune-riffle	Colluvial	Bedrock
Semi-mountain	High (floodplain is absent)	Straight	Aa, A						
		Meandering	Gc, G, Fb, F						
		Island-Braided	Db, D, Dc, Da						
	Medium (narrow floodplain)	Straight				Ba, B, Bc			
		Meandering				Ba, B, Bc			
		Island-Braided	Db, D, Dc, Da						
		Anabranching	D, Dc, Da						
	Low (wide floodplain)	Straight				Ca, C, Cc			
		Meandering				Eb,E			
		Island-Braided	Db, D, Dc, Da						
		Anabranching	D, Dc, Da						
Plain	High (floodplain is absent)	Straight	Bc						
		Meandering	Gc						
		Island-Braided	D, Dc						
	Medium (narrow floodplain)	Straight					B, Bc		
		Meandering					B, Bc		
		Island-Braided	D, Dc						
		Anabranching	Da						
	Low (wide floodplain)	Straight					C,Cc		
		Meandering					E,C,Cc		
		Island-Braided	D, Dc						
		Anabranching	Da						

APPLICATION

N.I. Makkaveev [Makkaveev, 1955] postulated, "Each classification may be worthwhile if it has applied sense." The truth of the classification is checked through its application for different practical needs. Here we observe the ways how the fluvial processes concept was imported into the practice. The main task is to determine certain correspondence of stream types for river use and management needs. Fluvial processes' classifications in the contemporary World is intended for finding solutions in global, regional, and local scales.

Among them, there are disaster reduction and environmental management. In general, the approach is determined through two objectives: assessment and prediction of channel evolution, and stream inventory. If the observer knows the channel types and its controls, he can use this information predicatively to evaluate the risk and sensitivity to disturbance.

For Channel Evolution

Recognizing and ordering of stream types provides insight into sensitivity of change. Stream system changes can be due to flow, sediment, or many of the interrelated variables that have produced the modern channel. Both Rosgen [Rosgen, 1994] and Makkaveev-Chalov [Makkaveev, Chalov, 1986] classifications were intended through past observations of adjustments to provide information on progressive stages of evolution. It has been fully confirmed by certain correlations between recognized channel types and various hydrological and geomorphologic events and variables. Bank erosion, bed shifts, avulsions, and human activities cause progressive stages in channel adjustment. These adjustments occur partially as a result of change in stream-flow magnitude and/or timing, sediment supply and/or size, confinement changes, human-induced disturbance, and vegetation changes. Observed changes in channel morphology over time were successfully expressed in terms of stream type changes.

Channel classification for channel evolution prediction presents a few procedures in consecutive order. Firstly, the quantitative and qualitative description of studied reaches and corresponding identification of channel control variables according to Table 1 is given. The second step includes stream and channel pattern type recognition. Finally, the morphology and stream dynamics, both with the large experience [Chalov, 2008], are analyzed.

Stream (fluvial processes) types serve as the background for channel evolution prediction. It is well-known that plain and semi-mountain rivers are probably more susceptible to change than mountain ones. Moreover, different stream powers and confinement ratios present various morphologic adjustments. The exact causes of changes are connected with given channel patterns. Usually, the most stable are straight channels. Their initial forming of straight channels are correlated with amounts of certain conditions (low stream-flow magnitude, low sediment supply, and similar width of channel and valley). Channel pattern shifts are caused by changes in any of the above-mentioned variables which could be easily observed and measured.

The evolution of meandering channel patterns could be analyzed on the base of its stages in channel adjustment. It is defined by the ratio between kinetic energy and hydraulic resistance at the bend of a stream: l/L and $r/h = 1.00$ (l – bend length, L – bend pitch, r – bend radius of curvature, h – bend deflection) [Makkaveev, 1955; Makkaveev, Chalov, 1986]. At the beginning of the meandering process ($l/L > 1.15$), the bend has the smooth segment form. When the ratio l/L increases to 1.60, the bend shows a steep segment form, and when $l/L >$ 1.60, it can be cut off, or it can obtain complicated loop-shaped form or sinusoid-shaped form. According to Knighton [Knighton, 1998], two main types of cutoff development are recognized: neck cutoffs and chute cutoffs.

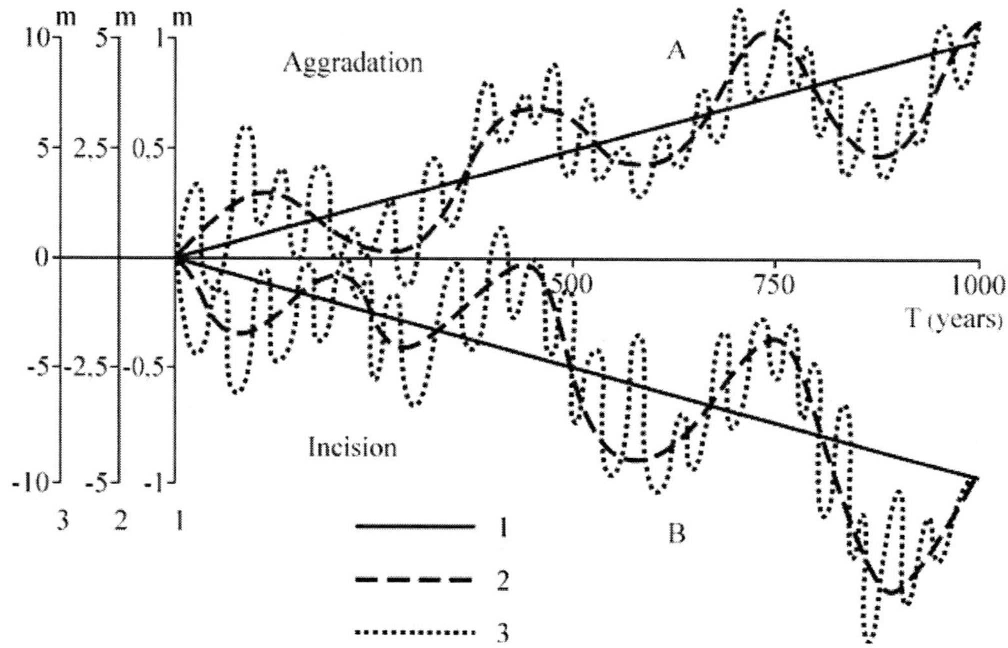

Figure 5. Superposition in time T of vertical channel changes at different scales: 1) geological timescale general directional; 2) medium-term periodical local, associated with channel planform evolution; 3) short-term alternating local, associated with bed forms migration. A – under general trend of channel aggradation, B – under general trend of channel incision. Each vertical axis (1, 2, 3) corresponds to channel bed elevation relative to conditional "0" for respective types of vertical shifts.

Concerning the intensity of meander development, certain relations of meander channel migration rate from the value of the ratio l/L [Makkaveev, Chalov, 1986] or from the bend curvature r_c/w, where r_c – radius of curvature, w – channel width [Knighton, 1998] were recognized. Hickin and Nanson [Hickin, Nanson, 1984] demonstrated that the rate of migration reaches a maximum when $2 < r_c/w < 3$, decreasing rapidly on either side of this range.

Different time scales of channel pattern shifts could be observed (Figure 5). In the geological time scale, these processes lead to changes of river longitudinal profiles and modifications of valley morphology. Continuous sediment deposition leads to channel aggradation and rise of water levels. This eventually results in burying older channel and floodplain alluvial deposits by modern sedimentation and the widening of a valley bottom induced by expansion of areas inundated during flood events towards terraces. In the historic time scale, effects of fluvial processes are noticeable when considering separated river reaches. The process is usually connected with channel pattern changes. Seasonal variability of channel patterns correspond to the sediment flux and flow transporting capacity inequalities. Those inequalities strongly differed between various stream types. Rapidly changing structures of a flow velocity field reflects a number of influences, such as intra-seasonal fluctuations of flow discharge and sediment transport or formation, development, and destruction of turbulent eddy currents. Resulting irregularity of a flow velocity field, in turn, induces the transversal non-uniformity of erosion and deposition, meaning that either scouring or settling of sediment particles prevails at different parts of a given cross-section of a river channel.

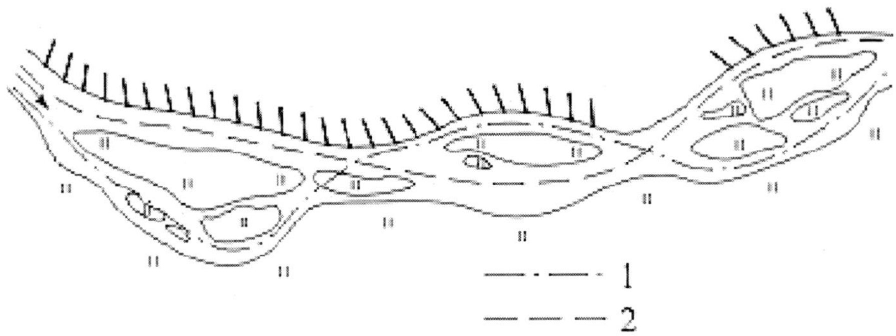

Figure 6. Evolution of conjugated braided reach of the Ob river. Location of the main water flow: 1 – at the beginning of the XX century, 2 – at present time.

Various types of braided channels require different approaches to channel regulation. Cyclic water redistribution and further morphological changes between branches of conjugated and parallel-braided braided patterns seem to be a key point for stream change predictions. It plays the most important role for braided channels of plain rivers. The "figure of eight" rule at conjugated braided channels means that the sequent water redistribution between the right and left branches with time intervals up to dozens of years at large plain rivers (Figure 6). In this and many other cases of navigation maintenance, this was the key task of fluvial processes studies in the USSR.

For Fluvial Management and Safety Maintenance

The same approach is used for hazard predictions. As it was mentioned above, channel development in a varying degree determines appearance of certain problems and river resource employment in different economic activities. Intensive channel evolution frequently provokes dangerous occurrences in channel processes and hazards for certain human facilities. In turn, channel training in various economic targets leads to changes of fluvial regime and channel process development and certain negative consequences.

Simulation for predicting river behavior is possible only when relationships between the governing factors are known. Accounting of spatial distribution of different channel types and peculiarities of their evolution is necessary while planning certain economic activities and prevention of negative and catastrophic sequences of channel processes. The true determination of the river type and corresponding dynamics is the base for proper and environmentally-based channel regulation, river channelization, managing the fluvial system, etc.

The whole variety of hazardous processes are provoked by one of the certain causes which could be classified either by spatial scale (regional or local) or origin (natural or human-induced). Thus, natural regional (water and sediment flow fluctuations), natural local (log jam formation), human-induced regional (mining, water transfers, dams and hydropower stations), and human-induced local (bridges, pipeline channel crossings) could be recognized as the main effects on fluvial systems. Meanwhile, they influence different elements of economical activities , situated in different parts of the river valley: in-channel objects (underwater crossings, etc.), near channel objects (dams, bank levees, wharfs), and floodplain objects (cities and settlements, etc.). River bank and bottom erosion, flooding of riverine areas, drying, and silting of river channels – are recognized to be the main forms of dangerous occurrences of channel processes [Berkovich, etc., 2000].

The following economic targets provoke the strongest influence on fluvial regime and channel processes and challenge the most serious approaches to channel processes investigation and regulation.

Meander's Cutoff

A well-known mode of channel management is artificial cutoff which is widely accepted along braided or sinuous reaches to provide either stream stability increases or water redistribution. Artificial channel cutoffs are widespread on the rivers with unstable channel regimes. Particularly, they take place on the reaches where certain centers of population or economic objects are situated on the bank under the strong influence of erosion processes – concave banks of meanders and unstable banks in braided channels. In these cases, due to construction of channel cutoffs, the stability of given reaches could be enhanced.

The vivid example recently comes from the town Kolpashevo which is located on the Ob river right bank. Intensive development of the meander provokes right bank erosion within the town (Figure 7). During the last 100 years, destruction of 1.5 km of the town's territory took place; the cemetery and the stadium were washed away. At present time, the meander continues its development and destructions of city buildings are expected; in 50 m near the contemporary location of the channel, a street connecting the upper and the lower parts of the town is situated. Through the channel evolution analyses and two-dimensional hydrodynamic modeling, the general scheme of channel cutoff construction to decrease the intensity of meander development and right bank erosion was developed (Figure 7). On the first stages, it was necessary to clarify the channel pattern types and its evolution in the XX century in order to integrate channel regulation modes to natural channel development. Thus, the exact determination of the channel types is of crucial importance for fluvial management.

Construction of Water Intake Systems and Water Outlets

Water supply engineering for communal services and industry include water intake systems and water outlet construction. These activities lead to either decreases or increases of water flow. On the other hand, channel processes development in certain cases provoke serious difficulties for the regular functioning of intake systems and systematic water supply of city areas or industrial facilities. First of all, this is connected with filling of water intake systems with sediments, shifts of side-bars towards water intakes, displacement of river bends when water intakes with the lapse of time turn out to be situated in meander cutoffs on the flood-plain, redistribution of water discharge between channel branches on braided rivers, and even the drying of them. Such situations took place on the braided reach of the Lena river near Yakutsk city: cyclical changes of water discharge redistribution between channel branches determines certain difficulties for the functioning of the Yakutskiy water intake system .

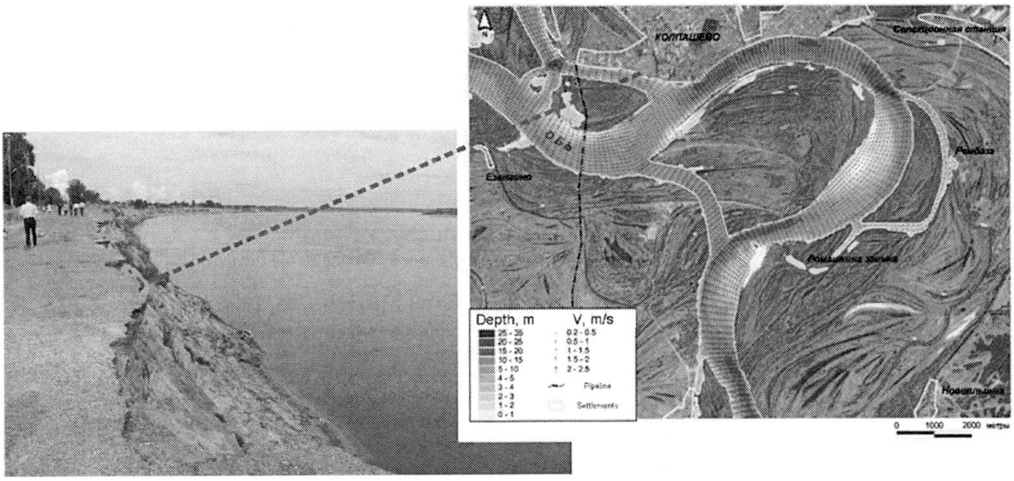

Figure 7. Bank erosion of the Ob river within the town Kolpashevo (left) and project of channel cutoff (right) to redistribute water.

Special concern for construction of sewage systems along braided reaches is also connected with special impacts of the water intake on the environmental flow in branches. The smaller size of the stream branches the larger impact of water intake on environmental flow in the given branch. The crucial decrease of water community states occurs when the water discharge rate is below some special value, as defined by Q_{cr}. To maintain a river branch in some agreed ecological condition could be done by exceeding Q_{cr}.. Amount of available water resources could be estimated as the difference between annual water discharge, Q_o, of non-braided reach and Q_{cr}, (usually equals 99% probability water discharge)

$$Q_{Д,i} = \phi_i \, (Q_o - Q_{i\,99\%}), \qquad (1)$$

where ϕ_i is relative water discharge.

Construction of River Crossings (Bridges, Pipelines, etc.)

The number of impacts due to river channel development is connected with different means of communication – bridges, pipelines, etc. So far as settlements are closely connected with river systems, road networks are located mostly along river valleys, so there are many bridges crossing them.

Geologic and geomorphologic factors along mountain rivers prevail channel development and further lead to the distribution of rather stable channels there. Frequency of bank erosion and the development of hazardous processes on these rivers is minimal, but only if there is no transformation of stream flow pattern under the influence of human activity. Otherwise, even low potential risks of channel processes development transform to actual danger. High stream power and constriction of effective cross-sections due to bridge piers are the main reasons for intensive bank erosion and destruction of communications (Figure 8,A).

Moreover, bridges are the cause of backwater spreading upstream , sediment accumulation, and formation of central bars and logjams exactly near their piers; this leads to intensification of river channel changes. Thus, on the Andrianovka river (the Kamchatka river basin), sediment accumulation and logjam formations near one of the bridge piers determined the main stream flow shift , and that provoked erosion of the latter and formation of emergency situation (Figure 8,B). In addition, bridges are one of the significant factors of island and logjam formation directly near their piers which also causes intensification of channel pattern shifts. The result is frequent destructions of crossing constructions (bridges, pipelines etc.).

Sufficient investigation before construction of the object is the main task for safe maintenance of the facilities. The prognosis of channel development, horizontal and vertical channel shifts, should be based on the assessments and comprehensive descriptions of the channel types. Various examples of the bridge construction prove these thoughts (Figure 8,C). Even in the initial stage of the bridge exploitation, serious mistakes in the prognosis of channel processes development becomes apparent. Such situations took place on the Ob river near Novosibirsk (Figure 8,D), where the bridge was considered to be located within the reach of free meanders on the stage of intensive evolution; concave banks of the bend are characterized by erosion processes of rather high intensity.

Figure 8. Destruction of bridge piers and bridges on the rivers of different types: A – mountain river, channel without developed alluvial form (the Saraynaya river, Kamchatka peninsula, 2005); B – semi-mountain river, anabranching channel (the Andrianovka river, Kamchatka peninsula, 2007); C – semi-mountain river, the tributary of the Kolima river, 2009); D – plain river, meandering channel (the bridge under construction on the Ob river, 2007).

Dams' Construction

Construction of various hydraulic structures on rivers usually leads to certain changes of water flow regime and channel processes, both upstream and downstream. Thus, construction of a dam in river channels provokes backwater formation upstream the dam. Mechanisms of channel processes in the zone of backwater spreading represents development of regressive sediment accumulation which gradually extends far from the boundaries of backwater itself. In certain cases, construction of dams and blocking of the river provokes silting and even degradation of the channel, considerable ecological damages, and decreases of water communities.

Downstream, the dams, due to decrease sediment load concentration and sediment flow erosion processes, develop; stream flow without bed load material possesses higher sediment-carrying capacity. Channel reach downstream is usually characterized by increases of channel width because bottom erosion does not compensate sediment load deficiency and stream transport capacity is realized particularly due to bank erosion [Berkovich, etc., 2000].

River Channelization

Special examples of the stream type effects in channel evolution come from zones of artificial channel construction. In the area of mining industry development, one of the most acute problems is the keeping of stream flows from the pollution. The problem may be solved by channel stream diversion construction around the mining area. Channel straightening turns out to be the main reason for bank erosion in channel diversions (Figure 9). The increase of channel slopes and transport capacity lead to the intensive incision of flow. Channel diversions are often located on the valley slopes far from the floodplain. Rapids and falls occur on the reaches where the channel is incised until the level of parent material. Rivers have an irregular longitudinal profile. Furthermore, the local areas of channel aggradation and incision appear.

Similarly to the needs of hydropower engineering construction of Hydropower Stations on channel diversions induse special changes in stream type evolution. In certain cases, construction of diversion stations provoke redistribution of water and sediment flow. Thus, on the Bistraya river (the Kamchatka river basin), the Bistrinskaya Hydropower Station was constructed on the channel diversion. But while constructing the Station and the channel diversion, the peculiarities of sediment flow regime were not taken into account. Consequently, a shift of bottom water sheets with high concentration of sediments occurred, which lead to its siltation and provoked serious disturbances of Hydropower Station functioning. Moreover, during winter periods, a great amount of sludge (brash ice) came into the channel, so that the station failed for the whole winter (up to 5–6 months).

For Fish Habitats

Fluvial processes vary in time and space along stream corridors and determine the quantity and quality of specific habitats in rivers. The distribution of the habitat patches (mosaics) change spatially over time due to primary drivers, particularly geomorphologic features of channel-flow interaction which changes in the direction from mountains to lowlands. The latest investigations show that the structure and production of river hydrocole communities depends on stream (fluvial processes) and channel patterns. The main distinctions occur at braided reaches. Even along short reaches, fish and benthos quantity increases at stream bifurcation.

Application of the fluvial processes theory in channel dynamic assessment and prediction and determining stream communities forming requires further development. Based on the data of Kamchatka peninsula rivers, the multiple regression analysis found a significant correlation between the amount of jams per channel length and types of fluvial processes FP (mountain, semi-mountain, plain), channel pattern type CP (straight, meandering, braided) (table 4), stream order *N*, slope *I,* and kinds of riparian vegetation:

$$Freq = 2{,}470 + 0{,}4206\, N + 0{,}0697\, I - 0{,}0650\, Veg - 1{,}0950\, FL + 0{,}4590\, CP, \qquad (2)$$

where FP and CP were measured as integral values (from 1 to 3 according to mentioned type in the same orders), and Veg – is the vegetation type (prevailance of chosenia and poplar; prevailance of chosenia; prevailance of willlow, but with presence of chosenia; distribution of willow and alder only).

Figure 9. The broken dam near the placer mining in Koryak region (June, 2008).

Table 4. Log jams at different channel patterns of Kamchatka peninsula rivers

Channel pattern	*Freq*, km^{-1} of channel length		Relative area $F_{jams}/F_{channel}$	
	Mean value	Variance	Mean value	Variance
Straight	38	0,85	0,039	0,036
Meandering	30,8	1,7	0,044	0,072
Braided	43,3	2,29	0,049	0,057

CONCLUSION

The needs of the practitioners demand the universal environmentally-based classification. It should be applied to each river and correspond to the burden of knowledge for the management and sustainability of the given fluvial system as the part of surface environment. We appreciate the contributions of many scientists to the theory of fluvial geomorphology.

State-of-the-art methods of our science requires new perspectives. According to continuum concepts, varieties of fluvial processes and forms represent a broad range of interrelated events, objects, and correlations. All of them are necessary for the channel evolution prediction and full stream inventory.

To gain these results, classification should be considered a framework of a much larger puzzle that primarily incorporates observations and laws of the given channel form control. The latter conclusion follows the practice demands and enhances the role of "criterion-free" classifications as a tool for observations and conceptual or mathematical modeling. It seems that the description of the events, controls, and morphological changes are the base of fluvial classifications.

REFERENCES

Alabyan A.M., Chalov R.S. Types of river channel patterns and their natural controls // *Earth surface processes and landforms,* vol. 2, 1998. Pp. 467–474.

Aleskeevsky N.I, Berkovich K. M., Chalov R.S. Erosion, sediment transportation and accumulation in rivers // *International Journal of Sediment Research,* 23, 2008. Pp. 93–105.

Berkovich K.M., Chalov R.S., Chernov A.V. *Ecological riverbed studies.* Moscow: GEOS, 2000. 332 p.

Bolisov S.I. *Biogenous relief formation.* Vol. 2: Zonality. Moscow: GEOS, 2007. 466 p.

Brice J.C., Blodgett J.C. Counter Measures for Hydraulic Problems at Bridges. Vol. 1. Analysis and Assessment. Report No. FHWA-RD-78-162, *Fed. Highway Admin.,* Washington, DC, 1978. 169 p.

Carson M. A. The meandering-braided river threshold: a reappraisal // *Journal of Hydrology,* 73, 1984. Pp. 315–334.

Chalov R.S. *Geographical research of channel processes. M.: the MSU Publishers,* 1979. 232 p.

Chalov R.S. Riverbed science: theory, geography, practice. Vol. 1. *Channel processes, factors, mechanisms, forms of manifestation and forming conditions.* M.: the LKI Publishers, 2008. 608 p.

Downs R.W. River Channel Classification for Channel Management Purposes. In: A. Gurnell and G. Petts (eds.), *Changing River Channels.* John Wiley and Sons, Chichester, 1995. Pp. 347-365.

Esin E.V., Chalov S.R. Logjams formation on the Kamchatka peninsula rivers and their role in distribution of fish fry // *Ecology,* (in print).

Ferguson R. I. The threshold between meandering and braiding // *Channels and Channel Control Structures,* 6, 1984. Pp. 15–29.

Gregory K.J. Vegetation and river channel process interactions. In: Boon P.J., Calow P., Petts G.E. (Eds.), *River Conservation and Management.* Wiley, Chichester, 1993. Pp. 255–269.

Gurnell A.M., Blackall T.D., Petts G.E. Characteristics of freshly deposited sand and finer sediments along an island-braided, gravel-bed river: The roles of water, wind and trees // *Geomorphology,* 99, 2008. Pp. 254–269.

Hassan M.A., Gottesfeld A.S, Montgomery D.R., Tunnicliffe J.F., Clarke G.C., Macdonald S.J. Salmon-driven bed load transport and bed morphology in mountain streams // *Geophysical research letters,* 2008, №38. P. 1-6

Hickin E.J., Nanson G.C. Lateral migration rates of river bends // *Journal of Hydraulic Engineering,* 110, 1984. Pp. 1557–1567.

Kellerhals R., Church M., Bray D.I. Classification and analysis of river processes // *Journal of the Hydraulics Division,* ASCE, 102, HY7, 1976. Pp. 813–829.

Knighton D. *Fluvial forms and processes.* Arnold, London, 1998. 383 p.

Kaszowski L., Krzemien K. Classification systems of mountain river channels // *River channels – pattern, structure and dynamics.* Cracow, Institute of Geography of the Jagiellonian University, 1999. Pp. 27–40.

Leopold L.B., Wolman M.G. River channel patterns: braided, meandering and straight // *Physiographic and hydraulic studies of rivers.* United States geological survey professional paper 282–B. Washington, 1957. Pp. 39–85.

Makkaveev N.I. *River channel and erosion in its basin.* M.: the Academy of Sciences Publishers, 1955. 347 p.

Makkaveev N.I., Chalov R.S. *Channel processes.* M.: the MSU Publishers, 1986. 264 p.

Makkaveev, N.I., Sakharova E. I., Chalov R.S. Contemporary eolian processes in the upper Ob' river valley // *Vestnik of the MSU. Ser. 5. Geography.* 1966. №2. Pp. 49–53.

Montgomery D.R., Buffington J.M. Channel-reach morphology in mountain drainage basins // *Geological Society of America Bulletin,* 1997. Pp. 596–611.

Naiman, R. J., Lonzarich, D. G., Beechie, T. J., and S. C. Ralph. General Principles of Classification and the Assessment of Conservation Potential in Rivers. In: Boon, R J., Carlow, R, and G. E. Petts (eds.), *River Conservation and Management,* John Wiley and Sons, Chichester,1992. Pp. 93-123.

Rhzanitcyn N.A. River channel forming processes. Leningrad: *Hydrometeoizdat,* 1985. 263 p.

Romashin V. V. Types of channel processes in relation to determining factors // *Proceedings of The State Hydrological Institute,* 155, 1968. Pp. 56–63.

Rosgen D. L. *A classification of natural rivers: Catena,* v. 22, 1994. Pp. 169–199.

Schaffernak F. Basics of River Morphology and Hydrotechnics. Vienna: Springer, 1950.

Schumm S. The *Fluvial System.* New York: Wiley, 1977. 338 p.

Schumm S. Patterns of alluvial rivers // Annual Review, *Earth and Planetary Sciences.* Vol. 13. 1985. Pp 5–27.

Smirnova V.G. Hydrologic morphometric analysis of braided rivers of the Altai region // *Ph.D. thesis.* Irkutsk: Siberian Department of Russian Academy of Sciences, 2002. 20 p.

Southerland W. B. *Stream geomorphology in glacial-fluvial valleys of the North Cascade Mountain range in Washington State* // Washington State University. Program in Environmental Science/Regional Planning and Department of Natural Resource Sciences. 2003. 141 p.

Van den Berg J.H. Prediction of alluvial channel pattern of perennial rivers // *Geomorphology* 12, 1995. Pp. 259–279.

Wolman M. G., Miller J. P. Magnitude and frequency of forces in geomorphic processes // *Journal of Geology,* 68, 1960. Pp. 54–74.

In: River Channels: Types, Dynamics and Changes
Editor: Daniel A. Molina, pp. 137-163 ISBN 978-1-61324-148-6

Chapter 4

RESPONSES TO RIVER MANAGEMENT? GEOMORPHIC CHANGE OVER DECADAL AND ANNUAL TIMESCALES IN TWO GRAVEL-BED RIVERS IN NEW ZEALAND

Ian C. Fuller*[1], *Jane M. Richardson*[1], *Les Basher*[2], *Rob C. Dykes*[1] *and Simon S. Vale*[1]

[1]Geography Programme, School of People, Environment & Planning, Massey University, Palmerston North, New Zealand
[2]Landcare Research, Nelson, New Zealand

ABSTRACT

River management in New Zealand's laterally active gravelly rivers has permitted floodplain development and protection of agricultural resources and infrastructure. Management of these dynamic systems has been hailed as a success for the approaches adopted, namely straightening and confining the river using bank protection. However, this activity also impacts river morphological / habitat diversity and potential gravel resource, by replacing broad riparian corridors with narrower channels and reduced lateral connectivity with the floodplain. This chapter quantifies river behaviour at a range of spatial and temporal scales by assessing geomorphic change in the lower Rangitikei River over 50 years and in the upper Motueka River over a 5 year period, using decadal aerial photo analysis and annual RTK-dGPS surveys respectively to address the effect of river management on channel morphology.

The lower Rangitikei has been transformed from a braided to predominantly single-thread wandering planform. This narrowed channel has remained relatively stable since 1983. Similarly, the upper Motueka has also been narrowed since 1960. Flood management practices such as river training by placement of willows and rock protection have stabilised these river reaches.

To quantify geomorphic change and associated sediment transfers more precisely in these narrowed channels, annual RTK-dGPS surveys between 2004 - 2010 acquired data to construct Digital Elevation Models (DEMs) in a ~1 km reach of the upper Motueka River. Morphological budgeting based on subtraction of successive DEM surfaces reveals complex spatial and temporal patterns of erosion and deposition. Overall, volumetric changes suggest these narrowed reaches have been net exporters of sediment, associated with continued channel degradation. This has left bar features, traditionally the focus of gravel extraction in the reaches, relatively isolated from all but extreme flows, limiting replenishment of the gravel resource.

INTRODUCTION

Alluvial river channel morphology is conditioned by sediment transport, erosion and deposition processes (Church, 2006). Gravel-bed river channels are naturally dynamic, adjusting to natural or anthropogenic changes in sediment supply or flow regime within the constraints imposed by boundary conditions (Lane and Richards, 1997; Brierley and Fryirs, 2005). The morphology of gravel-bed rivers is conditioned by discontinuous sediment transport conditioning processes such as bed and bank erosion (Couper and Maddock, 2001), which in turn effect, for example, riffle and pool formation (Booker *et al.*, 2001), bar development (Ashmore, 1991), meander bend initiation and migration (Gilvear *et al.*, 2000), chute cutoff and channel avulsion (Fuller *et al.*, 2003a). In gravel-bed rivers the coarse fraction of sediment, mobilised as bedload and constituting the bed and banks, is the major determinant of channel morphology (Leopold, 1992; Martin and Church, 1995; Church, 2006). Finer sediments are carried in suspension and, although they may represent a greater fraction of the total sediment load carried by a river, are less important in terms of channel adjustment in gravel-bed rivers (Martin and Church, 1995; Ham and Church, 2000).

Alterations in climate and flood regime can produce significant morphological adjustment in river channels (Martin and Johnson, 1987; Rumsby and Macklin, 1994; Werritty and Leys, 2001; Wishart *et al.*, 2008). Human activities can also have an impact on discharge or sediment transport processes that will drive morphological change in river channels (Gregory, 2006). Engineering works including channelisation (Williamson *et al.*, 1992; Simon and Rinaldi, 2006), dam construction (Kondolf, 1997), flow regulation (Shields *et al.*, 2000; Winterbottom, 2000) and gravel extraction (Rinaldi *et al.*, 2005; Wishart *et al.*, 2008) can directly affect river channel morphology. Anthropogenic actions can also influence river channels indirectly by altering sediment and flow through landuse change (Liébault and Piégay, 2002), including deforestation (Brooks and Brierley, 1997), urbanisation (Roberts, 1989) and agriculture (Knox, 1987; Gregory, 2006). Geomorphic adjustment may therefore be attributed to a range of forcing mechanisms (Rinaldi and Simon, 1998; Wishart *et al.*, 2008). For example, channel narrowing in the River Wear, northern England, was attributed by Wishart *et al.* (2008) to within-channel gravel extraction superimposed on a general trend of declining catchment sediment delivery and a twentieth century reduction in both flood frequency and magnitude. The diversity of natural river form along with a plethora of forcing mechanisms and process interactions make it difficult to predict the trajectory of river channel change (Downs, 1995; Hooke and Redmond, 1992). However case studies of fluvial systems that document spatial changes in channel form over time have the potential to identify some of the factors involved in geomorphic river adjustment and provide base line data for decision makers seeking an effective and sustainable approach to river management.

In some places, concern over the impact that human activity has had on aquatic environments has driven a shift in recent years from an engineering reach-based river management approach to a more holistic catchment ecosystem-based perspective with a focus on river rehabilitation (Brierley and Fryirs, 2008). Alternative approaches to engineering-based management are now receiving greater prominence as the consequences of this approach to river habitat and ecosystems are better understood. An ecosystem-based approach seeks to 'work with nature', enhancing and maintaining river system processes responsible for the dynamic nature of rivers, promoting adaptive and self adjusting fluvial systems (Palmer *et al.*, 2005; Piégay *et al.*, 2005). However in some cases the capacity for rivers to naturally adjust is limited by past engineering interventions that have essentially fixed and entrenched the channel, restricting lateral movement and disconnecting the floodplain (Spink *et al.*, 2009).

In the Rangitikei River, anecdotal evidence suggests considerable planform change in the latter part of the twentieth century, where replacement of a braided planform with a single, straightened channel occurred over the course of ~50 years. In the upper Motueka, the straightened, narrowed, channel is perceived to be degrading its bed. In this chapter we quantify channel planform change on three reaches of the lower Rangitikei River, New Zealand, using historical aerial photography and GIS analysis. Indices of channel change, including sinuosity, extent of braiding, rate of lateral shift and change in active channel width were used to identify the temporal and spatial nature of channel adjustments that have occurred between 1949 and 2007. We then look in greater temporal, but smaller spatial resolution at geomorphic changes between 2004 and 2010 in a c.1 km reach of the straightened upper Motueka River using detailed mapping of the riverbed to construct DEMs. DEM differencing permits quantification of sediment transfers within the reach using morphological budgeting, which provided a lower-bound estimate of annual sediment flux.

REGIONAL SETTINGS

The Rangitikei catchment drains 3933 km^2 and is located in the southern North Island of New Zealand (Figure 1). Catchment geology comprises greywacke in the headwater ranges, with weakly indurated mudstones and sandstones associated with the Wanganui Basin underlying the remainder of the catchment. Climate is temperate maritime, milder in the lowlands and colder in the central volcanic plateau environment in the north of the catchment. Average annual precipitation in the catchment ranges from 870 mm per year near the coast to over 2200 mm per year in the ranges (Horizons Regional Council, 2004). The mean annual flow of the river at Bulls is estimated at 74.68 m^3 sec^{-1}, based on the nearest gauge at Mangaweka (cf Figure 1) and catchment modelling (NIWA, 2010). Low-flow conditions dominate for most gauging sites between December and March (Horizons Regional Council, 2004).

The Motueka catchment drains 2076 km^2 and is located in the northern South Island (Figure 1). Catchment geology is more complex than the Rangitikei and includes basement igneous, ultramafic and sedimentary rocks, as well as extensive clay-bound gravels (Basher, 2003). The temperate maritime climate is similar to Rangitikei, but rainfall is higher, ranging from c.950 mm in the east of the catchment to >3500 mm in the western ranges (Basher, 2003). Mean annual flow in the upper Motueka is 9.24 m^3 sec^{-1} at the Three Beaches study site based on the combined flows recorded at the Motupiko and Motueka Gorge flow gauges (cf. Figure 1).

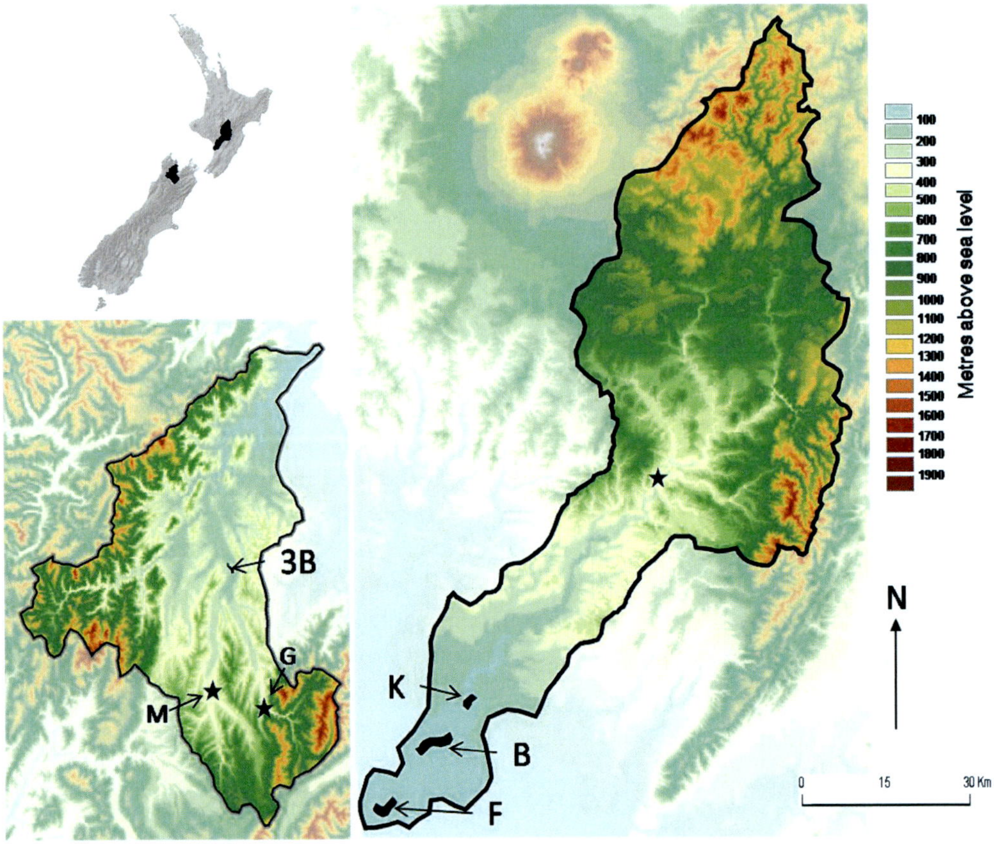

Figure 1. Location maps for the Rangitikei and Motueka catchments showing relative positions in New Zealand and catchment topography. Gauging sites referred to are marked with a star (M: Motupiko, G: Motueka Gorge). Study sites are labelled: 3B: Three Beaches, F: Flock House, B: Bulls, K: Kakariki.

River Management

Rangitikei River Scheme

The lower part of the Rangitikei River is managed under the Rangitikei River Scheme, primarily devised to control flooding and erosion. Flooding is controlled by 14.3 km of stopbanks (levees) designed to contain a 50-year return period flood. Stopbanks are confined to strategic locations, protecting settlements and bridges. Bank protection works include over 20 km of riparian trees, over 150 000 tonnes of rock work and an erosion protection reserve, comprising 1 115 ha of willow trees and 71 ha of pine plantation (Horizon Regional Council, 1999). The Rangitikei-Wanganui Catchment Board began the first Rangitikei River Scheme in 1952. These early river control works led to the reclamation of c.460 Ha of farmland as a result of bank protection works (Horizons Regional Council, 2000) and began to reduce the active channel width. River control works under the Rangitikei River Scheme No. 2 commenced in 1965 and were responsible for the construction of the majority of the scheme stopbanks and further bank stabilisation works (Horizons Regional Council, 2000). Up until the late 1970s, when rock work was more often used, bank erosion prevention was mainly achieved using protective planting and anchored trees (Horizons Regional Council, 2000).

Motueka Catchment Control Scheme

Management of the river for flood control purposes as part of the Motueka Catchment Control Scheme has resulted in significant modification of the main stem of the Motueka River and some of its tributaries (Basher, 2003). River control works (fairway clearance and bank protection) were implemented in 18 km of the upper Motueka in 1958. Complementing these river control measures were soil conservation farm plans and erosion control schemes, primarily for gully and streambank stabilisation on Moutere gravel terrain (Basher, 2003). In the upper Motueka, Basher (2003) lists the major works initiated in 1982 as part of the Motueka Catchment Control Scheme, including: stopbanks sited at strategic locations (settlements and bridges); establishment of clear channels of uniform width using a combination of river training, fairway clearance, bank protection (using rock, plant materials, or a combination of both), and groynes; provision of vegetation screens along riverbanks to contain the spread of water and sediment from rivers during floods. The major impact of river control has been to narrow and straighten the main channel (Basher, 2003).

Chapter Aims

In this chapter we use the lower Rangitikei as a case study of channel changes on a decadal timescale during a period of significant intervention in the river, which captures a dramatic change in channel planform. The planform of the upper Motueka has been less dramatically affected since inception of the catchment control scheme (Ball, 2004), nevertheless, we use a reach of the upper Motueka as a case study to assess in more detail morphological change and sediment transfers that take place in a narrowed, engineered, river reach. We use the upper Motueka because it is of a manageable scale and size, the size of the lower Rangitkei would preclude the same level of detailed ground survey as we deploy here. Thus the upper Motueka provides insight into processes operating in narrowed and straightened channels, such as the lower Rangitikei.

Method

Decadal Change: Rangitikei

Three reaches of the lower Rangitikei River were identified based on the availability of aerial photo coverage (Figure 1). The aerial photographs were scanned into TIFF® format and the data imported into ESRI ArcGIS® 9 Geographic Information System (GIS) for rectification and data processing. Aerial photographs were geo-referenced to New Zealand Map Grid co-ordinates in ArcMap™ 9.2 GIS by rectifying images to 2000 ortho-imagery of the lower Rangitikei. Each aerial photograph was rectified using 15 ground control points consisting of well spaced, easily identified buildings and road intersections. The positions of these ground control points were then used to calculate the Root Mean Square error (RMS). The RMS quantifies how consistent the transformation is between the different linked ground control points and provides a good assessment as to the accuracy of the transformation.

Each aerial photograph was rectified using either a 1st order affine or 2nd order polynomial transformation with a resultant RMS of below 6.5 m. The older, smaller scale aerial photographs produced lower RMS values when an affine transformation was performed, while the more recent, larger scale aerial images required a 2nd order polynomial transformation to produce a low RMS. The RMS errors determined for this study are comparable to those quoted by other authors who have used rectified aerial images to quantify channel change (Gurnell, 1997; Winterbottom, 2000). The RMS error suggests that positional changes greater than 6.5 m were likely to be the result of genuine movement rather than errors due to data handling.

Following rectification, aerial photographs were used in ArcMap™ 9.2 GIS as the basis for digitising boundaries of discrete morphological units comprising the active channel. A major problem when using aerial photos to identify these boundaries is the inconsistencies introduced to channel planform measurements as a result of difference in water levels on different survey dates. In order to avoid these inconsistencies this study adopted the definition of active channel boundary as described by Winterbottom (2000) and Wishart *et al.* (2008), which not only includes the wetted area, but considers unvegetated and sparsely vegetated gravel as being within the active channel zone. Lack of vegetation development suggests relatively regular inundation (once to several times per year) and indicates that semi-vegetated and bare gravel can be defined as part of the active bed. Areas completely covered in vegetation were only digitised as part of the active channel when surrounded by semi-vegetated gravel, active gravel or located within the wetted channel. It is possible that a single large magnitude flood could also clear vegetation, in which case the active channel would be deemed to have expanded to accommodate the high flow event.

Polygon digitisation and attribute allocation in ArcMap™ GIS using sequential aerial photographs for the three study reaches permitted the identification of not only the active channel boundary, but differentiated between vegetation cover, mid-channel and lateral bars during the period 1949-2007.

Indices for Assessing Channel Planform Change

Data collected from digitised aerial photographs were used to quantify a number of indices of channel planform change between successive dates. The digitised channel boundary was used to measure reach length at the channel midpoint, direct length between reach endpoints, length of mid-channel bars and channel width for each date between 1949 and 2007. Active channel width was measured as the shortest distance between channel margins at equally spaced cross-sections for each of the three reaches. Thirty cross-section positions were identified in the earliest photos and were spaced at 300 m intervals in the Bulls Bridge reach, 150 m in the Flock House reach, and 100 m apart in the Kakariki reach of the river. Where significant channel changes had occurred between aerial photograph dates, channel width measurements were made closest to the original cross-section position.

Stability of the study reaches over space and time was also investigated using the ratio of the active channel width to the total width of floodplain (or floodplain swath) occupied over the period of study. Floodplain swath was the maximum extent of the active channel across all surveys, derived by overlaying the active channel data from the total study period. Width was measured on each given date at equally spaced cross-sections. This ratio has been used to assess planform change by a number of authors including Gurnell *et al.* (1994) and Leys and Werritty (1999). A high ratio of active channel to floodplain swath indicates that there has been channel movement and instability over time (Leys and Werritty, 1999). A low ratio reflects the inability of the river to rework floodplain sediments and indicates channel stability over time (Leys and Werritty, 1999).

An additional indicator of channel behaviour is lateral shift, involving processes such as bank erosion, bend migration and channel avulsion (Leys and Werritty, 1999). In order to quantify lateral shift in the study reaches, total shift in m^2 between channel midpoints in successive surveys was calculated and this figure divided by length of the reach in metres. The results were reported in $m^2\ m^{-1}\ year^{-1}$, obtained by dividing the amount of lateral shift by the time between aerial photographs, a similar approach to that used by Werritty and Leys (2001). In order to calculate comparable rates of lateral shift throughout the period of study the time elapsed between aerial photographs should be the same. In this study the period between photographs was determined by the availability of suitable aerial photography, which meant that the length of time elapsed between aerial photographs ranged from 3 to 12 years. Lateral shift rates calculated from longer periods may fail to capture the full extent of channel movement occurring between survey dates.

Historical planform reconstructions from sequential aerial photos permitted definition of composition and nature of the different morphological units within the study reaches. Analysis of change in total active channel area over time provided some indication of channel behaviour, as only a significant change in boundary conditions or the crossing of a geomorphic threshold or impact of a large, high magnitude flood, has the potential to substantially alter the area occupied by the channel (Leys and Werritty, 1999). The composition of the active channel was also examined in order to provide some clarification as to the nature of change occurring within the active channel.

Planform transition trends between single thread behaviour and a braided morphology in river systems have often been detected and analysed using indices such as a braiding index and changes in the number and area of mid-channel bars (Passmore *et al.*, 1993; Winterbottom, 2000; Wishart *et al.*, 2008). In this investigation the number, area and length of mid-channel bars was used to calculate a braiding index, determine the extent of braiding within the study reach, and identify planform change over time. The braiding index was calculated as twice the total length of bars within the reach divided by the mid-channel length as defined by Brice (1960).

Channel sinuosity has often been used to examine channel behaviour, as changes in sinuosity can result from alteration in sediment or hydrological regime (Werritty and Ferguson, 1981; Leys and Werritty, 1999; Werritty and Leys, 2001). In this investigation reach sinuosity was calculated for each survey period by dividing the reach length at the channel midpoint by the straight line length between reach endpoints.

Annual Change: Motueka

Here we used a morphological budgeting approach to assess channel morphological change and sediment transfers responsible for such change using detailed ground survey in a c.1.5 km reach of the upper Motueka (cf. Figure 1). Survey data were used to construct successive DEMs of the reach.

Surveys

Successive annual surveys of the c.1.5 km reach of the Motueka River at Three Beaches (March or May 2004-2010) deployed RTK-dGPS and Total Station ground survey equipment. Two Trimble® R8 GPS receivers and two Leica® receiver units were deployed. One R8 unit acted as a base station, whilst the second R8 and two Leica® units acted as rovers. Data points acquired using this setup provide *x, y, z* coordinates determined on-the-fly using a 1-4 s occupation time per point (Brasington *et al.*, 2000; Fuller and Hutchinson, 2007). The use of RTK-dGPS allowed for rapid point topographic surveys to be conducted over a short period of time (Brasington *et al.*, 2000). The accuracy and precision of points acquired using this system have been measured by Fuller and Hutchinson (2007). They found an average horizontal precision of 0.006 m and an average vertical precision of 0.011 m when temporary benchmarks were resurveyed. A Topcon GTS 500 Total Station (± (5 + 5 ppm x D) mm, angular resolution 1 s) was used to acquire positions where satellite coverage was unavailable (e.g. beneath tree canopy at the margins of the channel, or at the base of cliffs). This was fully integrated with the RTK-dGPS survey. Each survey took five people approximately two days to complete under ideal operating conditions, i.e. low flows, and good satellite coverage enabling efficient function of the GPS.

The distribution of survey points collected throughout the reach was based on quasi-systematic point collection (Brasington *et al.*, 2000), where morphological units were used as a guide for point distribution. Complex morphology (e.g. slope breaks at bar and channel margins) were surveyed at higher point densities compared with units where surface changes were minimal and regular. Increased point density across areas of rapidly changing topography removed the need for artificial break lines to be placed in the data (Fuller and Hutchinson, 2007). This also increased the accuracy of detecting change between surveys of unstable areas of the active channel (Brasington *et al.*, 2000).

Table 1. Motueka Three Beaches survey data statistics

Survey	Area (m^2)	Survey points	Density pts m^{-2}	S (m)	ME (m)
March 2004	172746	4661	0.03	0.057	-0.005
May 2005	172642	8856	0.05	0.066	-0.004
May 2006	145143	10537	0.07	0.057	-0.003
March 2007	125087	10170	0.08	0.043	-0.003
March 2008	118123	11681	0.10	0.059	-0.004
March 2009	89882	9424	0.11	0.059	-0.003
March 2010	82169	11122	0.14	0.063	-0.002

DEMs

Generation

Surfer® spatial analysis software was used to generate DEMs from survey data. DEM interpolation used Triangulation with Linear Interpolation (TLI).TLI is based on a gridded version of the triangular irregular network (TIN), in which continuous triangular facets, irregularly sized and spaced, are created between data points (Fuller and Hutchinson, 2007). TLI connects original points so that a patchwork of triangular faces represents surface elevations. This method of interpolation works best for regularly spaced data, with original data points used as nodes for resulting triangulation (Surfer®, 2002). TLI therefore does not extrapolate z values past the data range, thus the resulting surface is closely related to original data and is a better representation of surface topography than other available interpolation methods such as kriging (Fuller and Hutchinson, 2007; Schwendel *et al.* in press). DEMs were based on a 1 m grid interval. The use of finer intervals can potentially introduce spurious artefacts into DEM generation (Brasington and Richards, 1998).

DEM Reliability

Quantitative error analysis of the interpolation method was performed using residuals. Residuals were calculated by using survey data points as control points and comparing these with the interpolated surface (Eqn. 1).

$$z_{res} = z_{surv} - z_{interp} \tag{1}$$

where z_{res} is the residual value, z_{surv} is the survey value and z_{interp} is the value of the interpolated surface.

Root mean square error is the most common descriptor of error in DEM analysis. Statistical analysis of residuals, using the approach recommended by Fisher and Tate (2006), allowed for the derivation of the standard deviation (*S*; Eqn. 2) and the estimation of bias using mean error (*ME*; Eqn. 3) associated with the interpolated surface for each DEM. Geostatistics produced in this error analysis for each DEM are shown in Table 1.

$$S = \sqrt{\frac{\Sigma\ [(z_{DEM} - z_{ref}) - ME]^2}{n-1}} \tag{2}$$

$$ME = \frac{\Sigma(z_{DEM} - z_{ref})}{n} \tag{3}$$

where z_{DEM} is the measurement of elevation form the DEM, and z_{ref} is the higher accuracy measurement of elevation for a sample of *n* points.

Note: point densities are below the spatial density of 0.69 m^{-2} and 1.10 m^{-2} attained by Brasington *et al.* (2000) in a similar study, although their reach area was significantly smaller. This data density is however consistent with similar work on equivalent scale reaches (e.g. Fuller *et al.*, 2003b; Fuller and Hutchinson, 2007), with point densities of 0.06 m^{-2} and 0.1 m^{-2} respectively. Ashmore and Church (1998) note that point densities of 3.5 m^{-2} recommended by Lane *et al.* (1994) produce unrealistic field survey problems (such as the time taken for data collection), and that point density should be tailored to the scale of topography which is being measured. A point density of approximately 0.1 m^{-2} was therefore considered acceptable due to the scale and topography of the reaches measured. Furthermore, reach average point density masks the underlying distribution of points throughout the reach and higher point densities at breaks in slope (Fuller *et al.*, 2003a).

ME values are very small and tend to be negatively biased. This means the interpolated surface on average lies slightly below the real surface by a matter of millimetres. The standard deviations do not exceed and are normally well below the roughness of the surface as defined by the D_{84} value of 110 mm (Fuller and Basher 2010).

DEM Differencing

Successive DEMs were subtracted from one another to create a DEM of difference, producing a surface representing the distribution of change between surveys. Change between surveys graphically represents three-dimensional changes in channel morphology. Furthermore, DEM differencing was also used to calculate net volumetric change within a reach. Volumes were calculated in Surfer® by finding the volume under a function $f(x, y)$ as defined in Eqn. 4.

$$\text{volume} = \int_{x_{\max}}^{X_{\max}} \int_{y_{\max}}^{y_{\max}} f(x,y)dxdy \qquad (4)$$

This is calculated by first integrating over X (columns) to obtain the areas under the individual rows, and then integrating over Y (rows) to obtain the final volume (Surfer®, 2002; Fuller *et al.*, 2005). Surfer® approximates the necessary one-dimensional integrals using three classical numerical algorithms (Surfer®, 2002). Net sediment volumes calculated in this study were based on the extended trapezoidal rule (after Fuller *et al.*, 2005).

Reliability of volumes generated: Since TLI does not provide an estimation of variance in its interpolation of surface elevations as other interpolation methods do (such as kriging), no precise quantification of accuracy was feasible. Fuller and Hutchinson (2007) suggested that volumes calculated using TLI were likely to be accurate to approximately ± 5%.

Detection of change: without surface variance, a precise quantification of minimum levels of change detection was not possible (see Fuller *et al.*, 2003b). However, for these purposes elevation changes which exceed 0.2 m were depicted in difference DEMs. This figure approximates twice the surface roughness of the reach as defined by D_{84}. Change < 0.2 m could simply reflect placement of survey poles on top of or at the base of gravel clasts which condition surface roughness.

Results

Rangitikei

Historic Planform Reconstruction

Historic planform reconstructions in the lower Rangitikei (Figure 2) show that in 1949 and/or 1955 the wide active channel was multi-threaded, comprising numerous bare gravel, vegetated and semi-vegetated mid-channel bars, high flow chute channels, abandoned anabranches and extensive lateral bar surfaces displaying various degrees of vegetation development. By 1983 all three reaches had narrowed considerably, displayed less morphological complexity and were predominantly laterally stable single thread reaches. The 2007 active channel almost entirely comprised active gravel surfaces, with relatively small isolated patches of semi-vegetated and vegetated lateral bars evident (Figure 2).

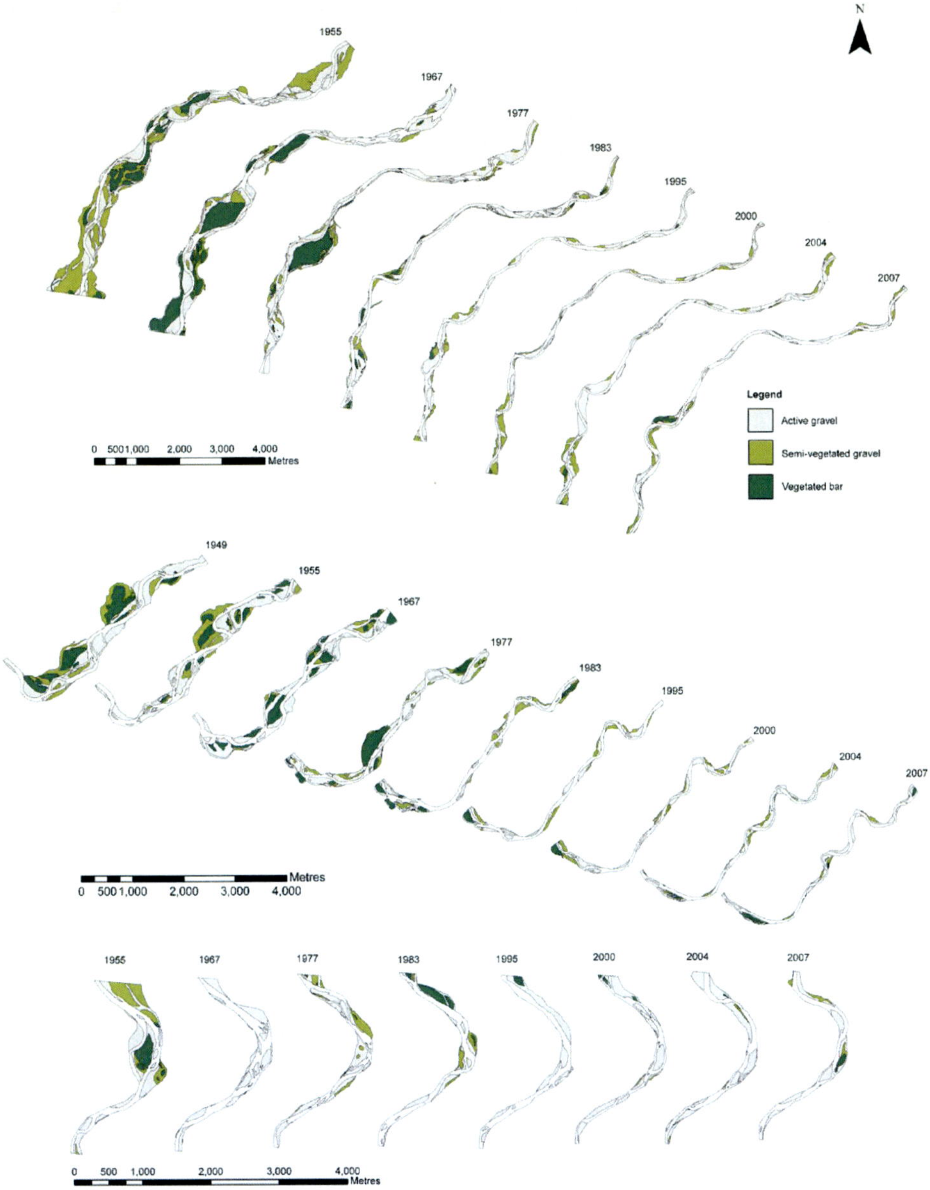

Figure 2. Channel planform changes at (a) Bulls Bridge (1955-2007); (b) Flock House (1949-2007); (c) Kakariki 1955-2007. Flow is from north to south in each.

Active Channel Width and Stability

The reduction in active channel widths derived from 30 cross sections in each reach is statistically significant (Figure 3). These results support the qualitative assessment of change derived from planform maps (Figure 2). Since 1983 the data suggest channel width has remained relatively stable with slight reductions in 2000 before an increase in 2004 (Figure 4). Overall between 1955 and 2007 the mean channel width of the Rangitikei River at Bulls Bridge reach has altered substantially, reducing by 74% (Figure 3). At Flock House mean active channel width reduced by 73% and at Kakairki a 48% reduction occurred. Reach stability was also assessed by calculating the ratio of active channel width to total active channel width (or floodplain swath) for the three study reaches (Figure 3). Channel width to floodplain swath ratios between 1955 and 1983 indicate a period of channel adjustment and narrowing before remaining relatively stable to 2007. At Kakariki most of the channel width change occurred earlier, between 1955 and 1967 with minor change since then. It is of note that the three sites have converged on the same channel width, which represents the original design width of the active channel (Figure 4b).

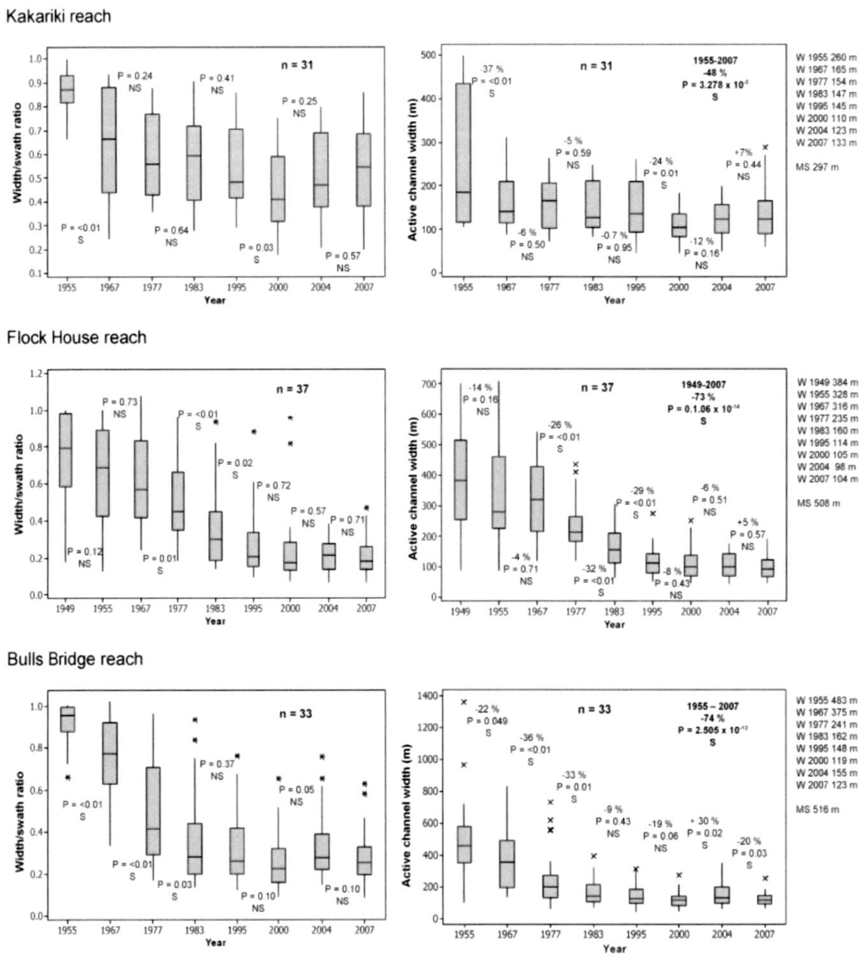

Figure 3. Rangitikei River active channel width and width/swath ratio changes in the study reaches between 1955 (1949 Flock House) and 2007; boxes represent inner and outer quartiles; vertical lines represent the upper limit = Q3 + 1.5(Q3-Q1) and lower limit = Q1-1.5(Q3-Q1); crosses are extreme values; n, number of width measurements; MS, mean value of floodplain swath; W, mean values of active channel width for each date; values in percent represents the rate of channel narrowing between dates; results of a parametric t-test are presented; S indicates significant differences of channel width and width/swath ratio between dates; NS, not significant.

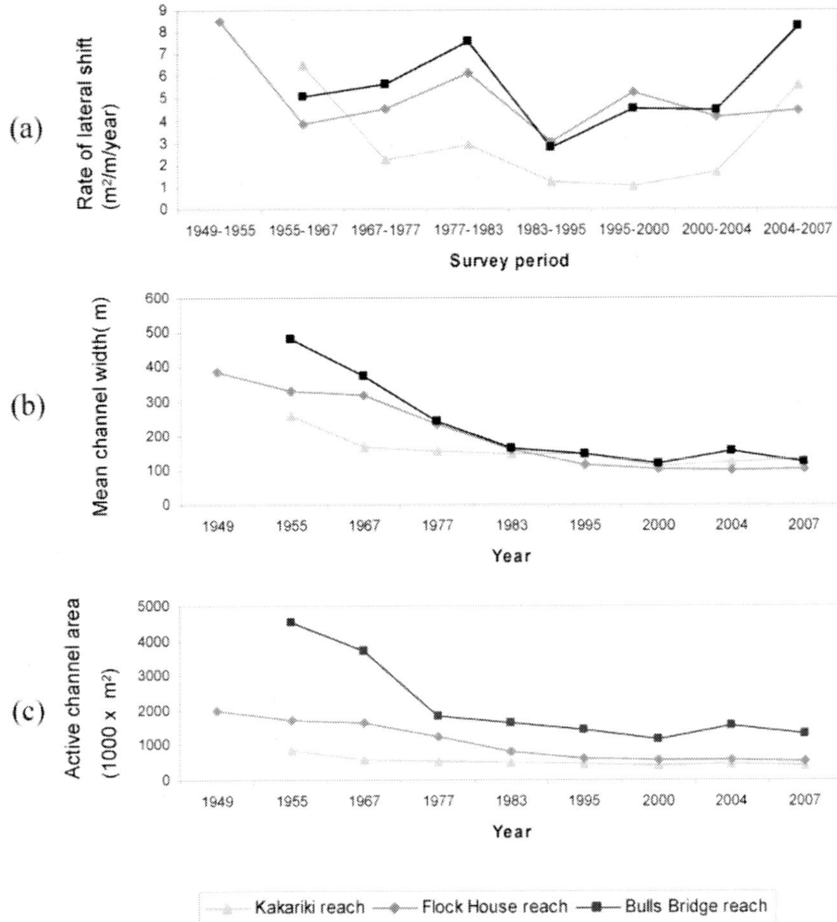

Figure 4. Changes in active channel parameters (a) rate of lateral shift (b) mean channel width and (c) active channel area, for the Kakariki, Flock House and bulls Bridges study reaches.

The rates of lateral channel shift calculated for each reach indicate that there has been variable movement of the channel midpoint (Figure 4a). Although direct comparison of the rates of lateral shift is made difficult by the unequal number of years between records it does give some indication of channel stability over the study duration. The Flock House reach was subject to the greatest rate of channel midpoint migration during the period 1949-1955 with a rate exceeding 8 $m^2 m^{-1} year^{-1}$ (Figure 4a). The Bulls Bridge reach experienced the greatest rate of lateral shift during two periods, 1977-1983 and 2000-2004. For the Kakariki reach the rate of lateral shift was highest between 1955-1967 and 2004-2007. The results illustrate similarities in the way Flock House and Bulls Bridge reaches have behaved over time, with variation in the rates of lateral shift mirroring each other between 1955 and 2004 (Figure 4a).

Active Channel Area

Quantitative analysis of active channel composition change was carried out for each study reach in the lower Rangitikei River (Figure 4c). Immediately evident is a marked reduction in the active channel area for all three reaches between 1949 and 1983, with earlier reduction and greater stability at Kakariki. Analysis of channel modification revealed that in less than 30 years the active channel area contracted by 64% at Bulls and 68% at Flock House (Figure 4c). Since 1983 there have been relatively small changes in the active channel area. There was a small (25%) expansion of the Bulls Bridge active channel between 2000 and 2004, but by 2007 the extent of the active channel had reduced again. At Flock House after 1995 very little variation in the total area occupied by the active channel is evident (Figure 4c). The consistently larger active channel area at Bulls Bridge is a product of longer reach length assessed in this study.

Change in Active Channel Composition

Quantitative analysis of the active channel composition in the study reaches highlights some aspects of the nature of channel change between 1949 and 2007. These results show that between 1955 and 1983 there was a marked decline in the area of active gravel and semi-vegetated gravel in the reaches (Figure 5). This is consistent with channel stabilisation, indicating less active lateral reworking of the channel in this period.

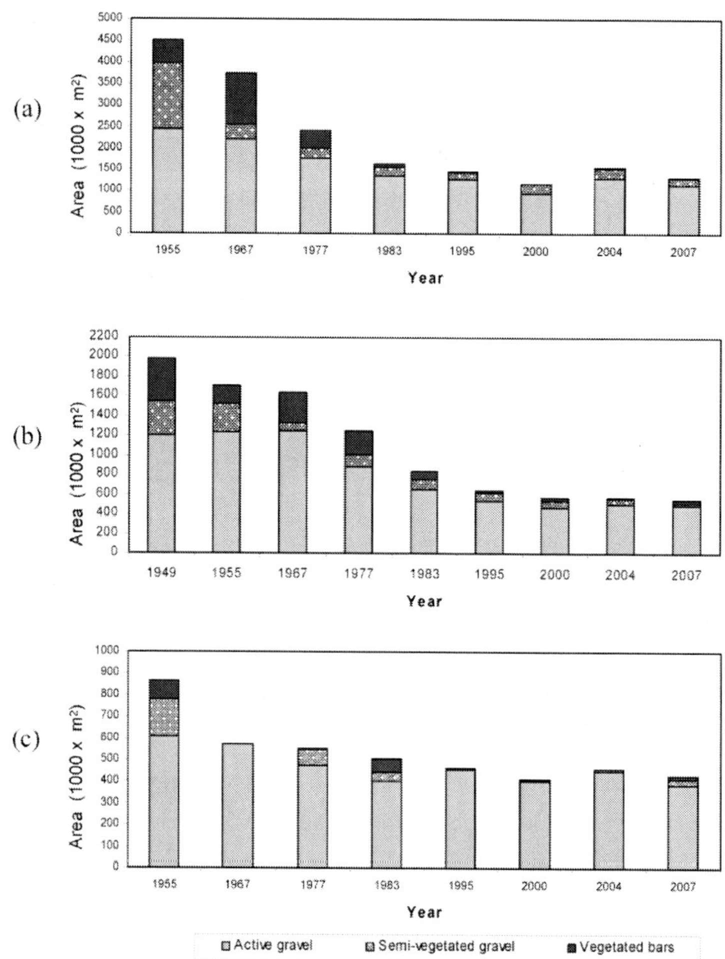

Figure 5. Changes in active channel composition showing active gravel area, semi-vegetated gravel and vegetated bars at: (a) Bulls Bridge, (b) Flock House, (c) Kakariki reaches.

Active channel geomorphology

GIS analysis of the active channel not only included the composition of the active channel but also differentiated between areas of wetted channel, lateral and mid-channel bars (Figure 6). Although the aerial photographs were taken during periods of low flow there will be some fluctuation in the relative proportion of bar type and wetted channel area. Despite these limitations some clear trends can be identified, with a reduction in the extent of lateral and mid-channel bars between 1955 and 1983 clearly visible. For example, in 1955 30% of the active channel area at Bulls Bridge comprised mid-channel bars, by 1983 this reduced to 5%.

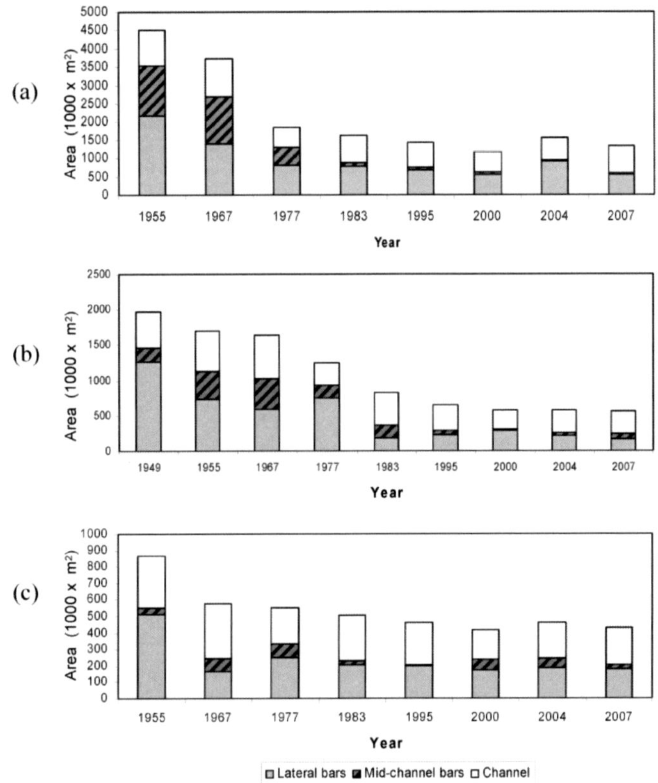

Figure 6. Changes in active channel geomorphology showing mid-channel and lateral bar area at: (a) Bulls Bridge, (b) Flock House, (c) Kakariki.

Braiding and Sinuosity

Figure 7 shows that not only has there been a reduction in the area associated with mid-channel bars between 1955 and 1983 but that the actual number of bars has also decreased. Leys and Werritty (1999) used an analysis of total medial bar area and number of bars to identify a trend from braided to single thread planform. This study uses a similar approach which identifies a reduction in braiding with fewer mid-channel bars detected within the study reaches and an associated reduction in the area of mid-channel bars between 1955 and 1983 (Figure 7). Whilst Bulls Bridge and Flock House show similar trends (Figure 7a-b), results from Kakariki do not show a clear trend (Figure 7c), although there is some indication that the reach was more braided in 1967 and 1977. A braiding index summarises these changes (Figure 8a), showing a clear reduction in braiding over the period of study.

Commensurate with this reduction in braiding is a slight overall increase in reach sinuosity, (Figure 8b).

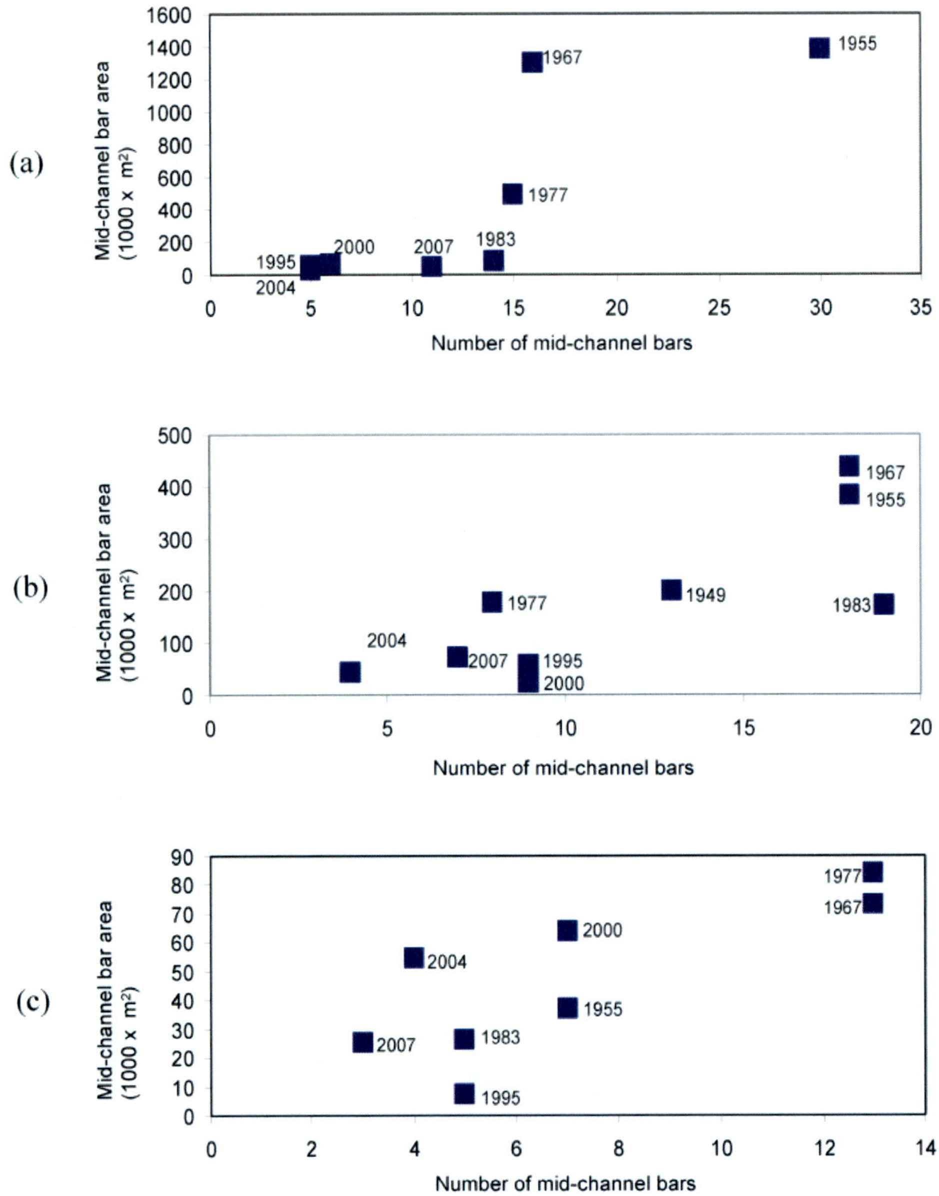

Figure 7. Changes in the number and area of mid-channel bars 1955 (1949 Flock House) -2007 at: (a) Bulls Bridge, (b) Flock House, (c) Kakariki.

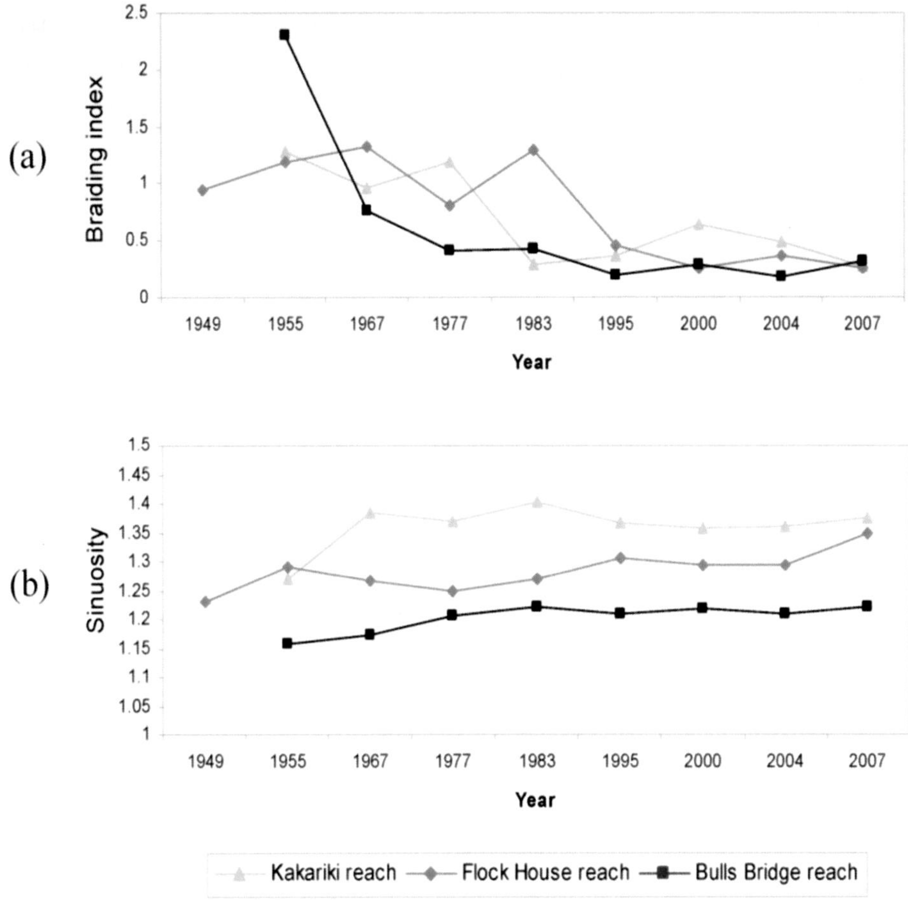

Figure 8. Changes in (a) braiding index and (b) sinuosity for the Bulls Bridge, Flock House and Kakariki reaches of the lower Rangitikei River 1955 (1949 Flock House)-2007.

Motueka

Figure 9 displays the DEMs derived from each survey and DEMs of difference depicting spatial patterns of morphological change between successive survey dates. Table 2 summarises sediment transfer volumes (erosion, deposition and net transfer within the reach). Immediately apparent in the DEMs (Figure 9a) is a narrowing of area surveyed over time. This represents a focus of each survey on the active surfaces in the reach. Abandoned / inactive surfaces at the margins of the active channel were not re-surveyed. The extent of surface activity was determined in the field based on inspection of debris, sediment and vegetation within the active channel. Initial surveys mapped all surfaces within the active channel. Subsequently, all bare gravel surfaces were included in each survey, but higher vegetated surfaces without any evidence of disturbance were not re-surveyed. Since morphological budgeting is premised on movement of bedload, unless new deposits of gravel or areas of erosion (devoid of vegetation) were evident, no change would be detected. Drapes of fine sediment were at times detected on higher surfaces, as was evidence of inundation (bent vegetation, woody debris), however fine drapes are not detectable by morphological budgeting and inundation without disturbance of the gravelly substrate does not contribute to a morphological budget either. The substantial amount of change detected between March 2004-May 2005 (Figure 9b and Table 2) is attributed to a 50 year flood which occurred in March 2005 (Figure 10).

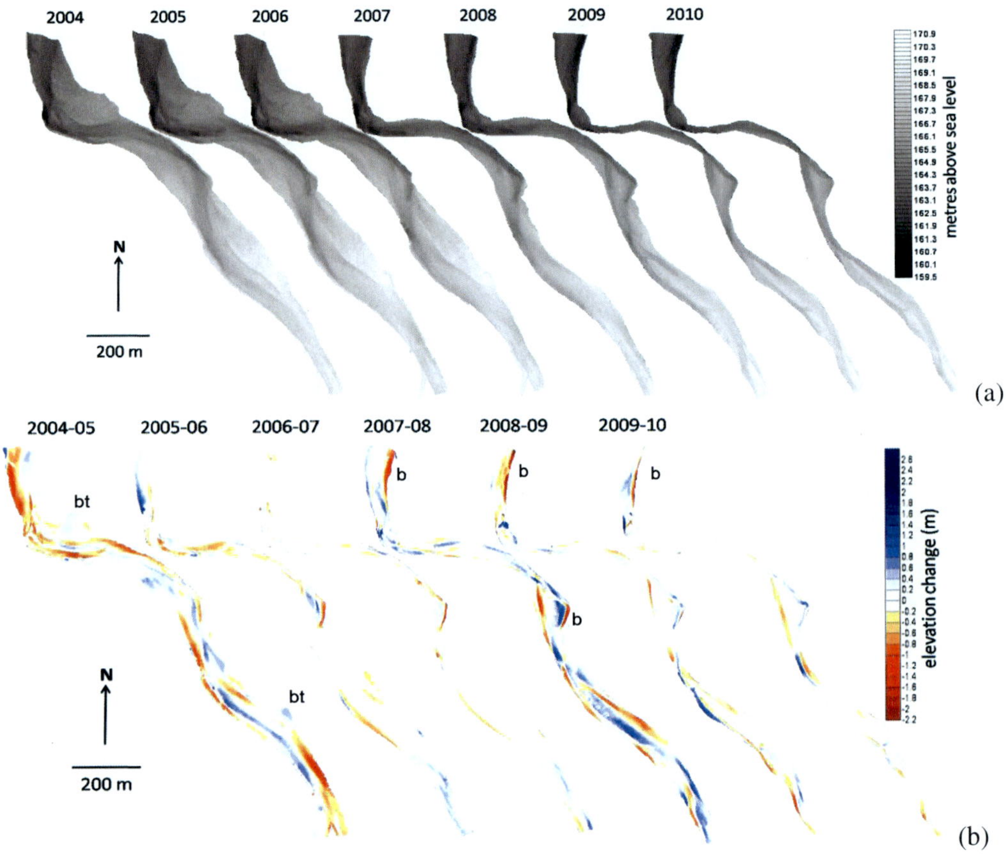

Figure 9. (a) DEMs of Three Beaches site, upper Motueka River, 2004-2010, (b) DEMs of difference: Three Beaches, upper Motueka River 2004-2010, 'b' indicates bank erosion, all remaining negative elevation changes reflect bed scour; 'bt' indicates bar top deposition.

Table 2. Sediment transfers: Three Beaches

Date	Erosion (m^3)	Deposition (m^3)	Net (m^3)
March 2004 – May 2005	27765	16418	-11347
May 2005 - 2006	9832	9043	-789
May 2006 - March 2007	7063	5573	-1490
March 2007 - 2008	14331	21945	+7614
March 2008 - 2009	10940	7721	-3219
March 2009 - 2010	7640	6800	-840
March 2004 - 2010	30377	20052	-10325

Figure 9b shows complex patterns of erosion and deposition within the reach and variable magnitudes of morphological change between dates. Bed level changes up to ±2 m occurred within a single year. Erosion typically occurred within the wetted channel and outside of bends with hard banks (rip rap) and deposition typically occurred within the channel as accommodation space is provided with lateral migration of the channel or with the stalling of bedload sheets across low bar platforms within the active channel (Plate 1).

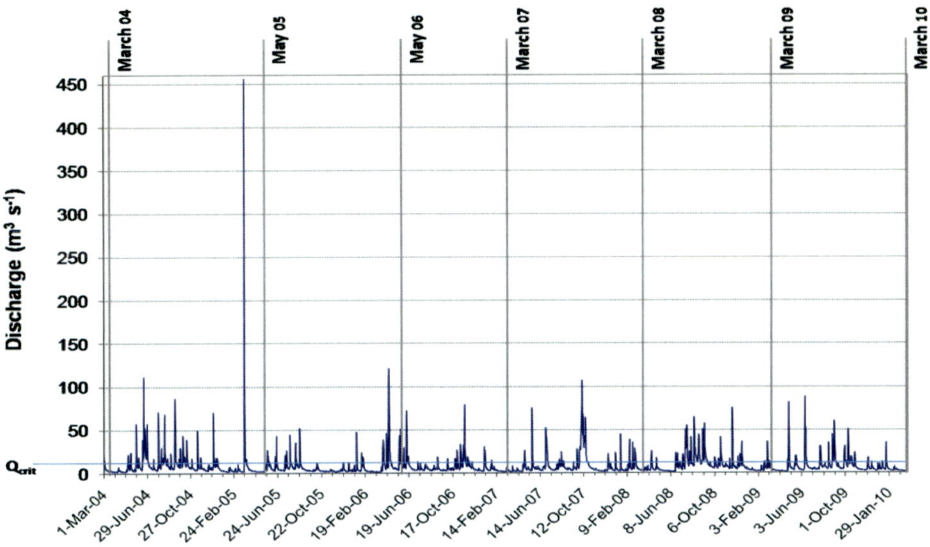

Figure 10. Motueka daily mean flow at Three Beaches. Derivation of critical discharge is described in Fuller and Basher (2010).

Plate 1. Stalled sheet of gravel and sand deposited across a stable, vegetating low elevation bar platform, Three Beaches, March 2008. The lower section of a survey pole can be seen in the mid-left of the image. The sheet is approximately 0.5 m deep. Flow is from left to right.

Discussion

Analysis of channel planform and associated indices quantifying change in planform over time in the lower Rangitikei has highlighted significant change in form between 1949 and 2007, notably channel narrowing, and reduction in morphological complexity, lateral activity and extent of braiding. Closer examination of change in channel planform indices over the study period reveals that the rate of active channel modification has not been constant over the period. The majority of changes in these indices occurred in the Bulls Bridge and Flock House reaches by 1983 (Figures 4-8). Since that time the active channel has remained relatively stable, with only small magnitude changes identified between 1983 and 2007. Evidence of channel change in the Kakariki reach suggests that most of the channel modification identified in the reach had in fact occurred by 1967 and since that time there has only been relatively minor change in the active channel (Figures 4-8). This suggests the measures used to constrain this reach were immediately effective and of the three reaches this was initially the narrowest, which also implies a naturally more stable reach.

Numerous studies, including investigations into channel change in mainland Europe (Marston *et al.*, 1995; Bravard *et al.*, 1997; Liébault and Piégay, 2001; 2002, Liébault *et al.*, 2002) and the British Isles (McEwen, 1989; Gurnell, 1997; Winterbottom, 2000; Wishart *et al.*, 2008) have identified similar changes in gravel-bed river morphology. Essentially many rivers underwent a transition from a wide, unstable braided form to a narrower and laterally (not necessarily vertically) more stable single thread channel. Channel change in these cases was conditioned by a number of factors including climatic changes and the onset of river engineering, with many of the potential factors coinciding making it difficult to differentiate between the individual effects of potential drivers of change. A review of the Rangitikei scheme in 1983 found that channel degradation, caused by channel confinement and gravel extraction, was leading to undermining and failure of tree bank protection works. The 1983 scheme review concluded that the 300-400 m wide channel design specified by Scheme Reviews 1 and 2 was too narrow. It is of note that the active channel widths in this study are even narrower. This is likely to be a consequence of attempts to keep the channel well within this 300-400 m wide fairway via straightening, such that the margins of the fairway stabilise, producing the much narrower channel observed. The Rangitikei River Consolidation Scheme commenced in 1985 with a number of proposals for the future management of the river. These proposals included widening the central channel fairway to 500 m, the use of rock work instead of tree protection for preventing bank erosion and the movement of gravel extraction sites from areas of degradation to aggrading reaches. In order to prioritise erosion control repair works, a riparian zone was delineated by the Consolidation Scheme, identifying areas where some risk of erosion would be acceptable. An extensive review of the Rangitikei River Scheme was undertaken in 1993/1994 and was initiated in response to concerns over the effects of degradation and aggradation on scheme maintenance costs. In addition, costly maintenance and removal of government subsidies in the late 1980s necessitated a consideration of the ways in which scheme expenditure could be reduced. The purpose of the review was to provide for the management of the Rangitikei River Scheme and the sustainable management of the gravel resources. The final report of the Rangitikei River Scheme Review (No. 3) adopted the River Management Programme for the Rangitikei River (Horizons Regional Council, 2000). This allowed for the prioritisation of river erosion protection by defining a 'Desirable Erosion Limit' (Figure 2). Included under the rules for prioritising river erosion protection was the restriction on placing rock protection in such a way that reduces the river meander zone to no less than 200 m width and ideally retains a meander zone of 400-500 m (Horizons Regional Council, 2000). The success of these rules is somewhat debatable, given the evidence for a far narrower active channel presented here (Figures 3 and 4), hence a fourth review of the Rangitikei river scheme is currently underway. One of the major issues facing the Regional Council is the costly maintenance of the scheme in its present form (pers. comm., Warren Wheeler, HRC).

The geomorphic changes detected in the lower Rangitikei over the past ~50 years may therefore be attributed to the way the river has been managed in these reaches. Limiting active channel width, confining flood flows and extracting gravel above the natural rate of supply (Richardson and Fuller, 2010) has resulted in a narrow, entrenched river channel, which is narrower than the river management scheme intended. The enforced stability of this channel is highlighted by the limited nature of geomorphic response to a flood in 2004 with an ARI of 50 years. This event barely registers in the evidence in Figures 2-8, with a small, but significant increase in width at Bulls being the most substantial change recorded. A flood event of similar magnitude in the Pohangina and Oroua Rivers, located within the adjacent Manawatu catchment resulted in more substantial change (cf. Fuller, 2007) compared with the Rangitikei. The Pohangina and Oroua Rivers are also both managed under a River Scheme, however, fairway widths are not as restricted relative to channel size and these rivers have very little rock protection and few stopbanks in comparison. The Rangitikei is therefore unnaturally robust in its response to flooding, which is a product of high levels of bank protection.

Processes operating in such narrowed channels can be elucidated using the more detailed investigation in the upper Motueka. Figure 9 and Table 2 demonstrate an overall degradational trend in the upper Motueka at Three Beaches. The one exception to net loss occurs between 2007 and 2008, when substantial net deposition occurred. This suggests net import of material into the reach from upstream during this period. The subsequent return to net degradation suggests that this response is probably a product of the passage of a wave of sediment through the reach at this time (cf. Wathen and Hoey, 1998), which is possibly in response to an earlier large (ARI 50 years) flood in the catchment in March 2005. DEM differencing provides an effective means of determining the morphological sediment budget as well as identify the complex patterns of sediment erosion and deposition occurring in the upper Motueka River. The results show that between 2004 and 2010, over 10,000 m^3 of sediment has been lost from the reach (Table 2). Losses in the reach were mainly confined to bed scour of the wetted channel with small areas of bank erosion contributing to the overall sediment deficit (Figure 9b). Rock armouring of banks and extensive willow plantings significantly limit the lateral movement of the channel at this site and have clearly served to reduce bank erosion during this period. The bank erosion which was observed ('b' in Figure 9b) took place not into floodplain *per se*, but vegetating, stable bar surfaces within the reach which have historically been the focus of gravel extraction.

Since the channel is laterally constrained, excess stream power is used to scour its bed, deepening the channel and generating the degradation evident. The result is self perpetuating, since as the channel deepens, higher flows become confined to within-channel and since the banks are armoured, further scour ensues. This may explain the persistent degradation trend in this reach between 2004 and 2010. However, other rivers in the Tasman District are also degrading, as are beaches, suggesting a larger scale reduction in sediment supply within the catchments may also contribute to degradation observed (cf. Wishart *et al.*, 2008). Channel incision has been shown to decouple channel and floodplain processes and disconnect large areas of a catchment from the sediment cascade (Fryirs *et al.*, 2007). In containing larger floods and concentrating flows, incised channels potentially generate higher transport capacities, resulting in the mobilisation of large volumes of bedload and further bed degradation (Fryirs and Brierley 2001). Furthermore, the absence of deposition on the higher bar platforms in the reach is striking, but hardly surprising given the short-term measurements and flood pattern. Only between 2004 and 2005 is any bar-top deposition recorded (Figure 9b) and then, only at a small scale. This period coincides with the largest flood on record in the reach (Figure 10). This may have serious implications on the availability of the gravel resource in this catchment, since Tasman District Council has previously extracted gravel from the reach by bar-top skimming, as it suggests the resource is not being replenished. Furthermore, bed degradation may also undercut and destabilise bank protection works.

CONCLUSIONS

Gravelly wandering rivers have been laterally constrained through management practices in the last ~50 years in New Zealand. The Rangitikei and Motueka are but two examples of typical practice. This confinement limits morphological diversity in such rivers. Channel constriction has also facilitated channel degradation and lateral connectivity between higher-relief bar platforms within the active channel and the wetted channel, as well as channel-floodplain connectivity, has been reduced. This has left bar features, traditionally the focus of gravel extraction in such rivers, relatively isolated from all but extreme flows, limiting replenishment of the gravel resource and potentially undermining expensive bank protection works. If communities consider these effects to be undesirable, a focus of river rehabilitation around the broader river corridor, rather than the river channel may allow for greater channel migration and adjustment to instability. The current Rangitikei scheme review is proposing removal of some bank protection because it is becoming too expensive to maintain bank protection structures in over-narrowed degrading reaches (C. Jordan, pers. comm.). This will reinstate channel-floodplain connections and sediment stored in the floodplain can only reach the channel network if the floodplain is reworked (Brierley and Fryirs, 2005). However, reach-to-reach connections from upstream may also be responsible for contributing sediment to these reaches, as observed in the Motueka between 2007-2008, although this supply may be readily interrupted by gravel extraction and the passage of such bed waves may take many years (Wathen and Hoey, 1998). Bank protection prevents bank erosion and lateral migration of the channel, stabilising the channel location, but reducing local sediment supply. Increasing the fairway width and allowing the channel to migrate more freely would enhance local channel-floodplain connectivity, replenish sediment without relying on upstream catchment inputs alone and enable the river system to adjust naturally to disturbance as well as improve the geomorphic diversity in the river. It is, however, fair to say that to allow rivers to adjust in this capacity may require a wholesale change in the way that rivers and adjacent floodplains are managed, which is undoubtedly a matter for society at large to consider.

ACKNOWLEDGEMENTS

The following are thanked for assistance with survey in the Motueka River: David Feek, Alastair Clement, Erin Hutchinson, Emma Phillips, Arved Schwendel (Massey University); Brenda Rosser (when at Landcare Research, now GNS Science); Christine Daly, George Knights, Stephanie Merrill, Maria Borlinghaus, Diarmuid Neilan & Claire Butty (Landcare Internships). Fish & Game (Wellington) provided assistance to JR to complete the work on the Rangitikei.

REFERENCES

Ashmore, P.E. (1991). How do gravel-bed rivers braid? *Canadian Journal of Earth Sciences*, 28, 326-341.

Ashmore, P. & Church, M. (1998). Sediment transport and river morphology: A paradigm for study. In Klingeman, P., Beschta, R., Komar, P. & Bradley, J. (Eds), *Gravel-Bed Rivers in the Environment* (pp. 115–48). Colorado: Water Resources Publications LLC.

Ball, L.G.A. (2004). *Quantifying the morphological effects of gravel extraction on the Motueka River*. Unpublished Master of Applied Science dissertation, Lincoln University.

Basher, L. (2003). *The Motueka and Riwaka Catchments: A technical report summarising the present state of knowledge of the catchments, management issues and research needs for integrated catchment management*. Landcare Research New Zealand, Lincoln, Canterbury.

Booker, D.J., Sear, D.A. & Payne, A.J. (2001). Modelling three dimensional flow structures and patterns of boundary shear stress in a natural pool-riffle sequence. *Earth Surface Processes and Landforms*, 26, 553-576.

Brasington, J. & Richards, K.S. (1998). Interactions between model predictions, parameters and DTM scales for TOPMODEL. *Computers and Geosciences*, 24, 299-314.

Brasington, J., Rumsby, B. & McVey, R. (2000). Monitoring and modelling morphological change in a braided gravel-bed river using high resolution GPS-based survey. *Earth Surface Processes and Landforms,* 25, 973–990.

Bravard, J., Amoros, C., Pautou, G., Bornette, G., Bournaud, M., Creuzé Des Chatelliers, M., Gilbert, J., Peiry, J., Perrin, J. & Tachet, H. (1997). River incision in south-east France: Morphological phenomena and ecological effects. *Regulated Rivers: Research and Management*, 13, 75-90.

Brice, J.C. (1960). Index for description of channel braiding. *Geological Society of America Bulletin*, 85, 581-586.

Brierley, G.J. & Fryirs, K.A. (2005). *Geomorphology and river management applications of the river styles framework*. Blackwell Publishing, Oxford, UK.

Brierley, G.J. & Fryirs, K.A. (2008). River *Futures: An Integrative Scientific Approach to River Repair*. Island Press, Washington DC.

Brooks, A.P. & Brierley, G.J. (1997). Geomorphic responses of lower Bega River To catchment disturbance. *Geomorphology*, 18, 291-304.

Church, M. (2006). Bed material transport and the morphology of alluvial river channels. *Annual Review of Earth and Planetary Science*, 34, 325-354.

Couper, P.R. & Maddock, I.P. (2001). Subaerial riverbank erosion processes and their interaction with other bank erosion mechanisms on the River Arrow, Warwickshire, UK. *Earth Surface Processes and Landforms*, 26, 631-646.

Downs, P.W. (1995). Estimating the probability of river channel adjustment. *Earth Surface Processes and Landforms*, 20, 687-705.

Fisher, P. & Tate, N. (2006). Causes and consequences of error in digital elevation models. *Progress in Physical Geography,* 30, 467–89.

Fryirs, K.A., Brierley, G.J., Preston, N.J. & Kasai, M. (2007). Buffers, barriers and blankets: The (dis)connectivity of catchment-scale sediment cascades. *Catena*, 70, 49-67.

Fryirs, K.A. & Brierley, G.J. (2001). Variability in sediment delivery and storage along river courses in Bega catchment, NSW, Australia: implications for geomorphic river recovery. *Geomorphology*, 38, 237-265.

Fuller, I.C. (2007). Geomorphic work during a "150-year storm: contrasting behaviours of river channels in a New Zealand catchment. *Annals of the Association of American Geographers*, 97, 665-676.

Fuller, I.C. & Hutchinson, E. (2007). Sediment flux in a small gravel-bed stream: Response to channel remediation works. *New Zealand Geographer,* 63, 169–180.

Fuller, I.C. & Basher, L. (2010). *Sediment transfers in the upper Motueka River: morphological budgeting 2004-2010.* Report prepared for Landcare Research, Massey University, 47pp.

Fuller, I.C., Large, A.R.G. & Milan, D.J. (2003a). Quantifying channel development and sediment transfer following chute cutoff in a wandering gravel-bed river. *Geomorphology*, 54, 307-323.

Fuller, I.C., Large, A.R.G., Charlton, M., Heritage, G.L. & Milan, D.J. (2003b). Reach-scale sediment transfers: an evaluation of two morphological budgeting approaches. *Earth Surface Processes & Landforms*, 28, 889-903.

Fuller, I.C., Large, A.R.G., Heritage, G.L., Milan, D.J. & Charlton, M.E. (2005). Derivation of reach-scale sediment transfers in the River Coquet, Northumberland, UK. *In* Blum, M., Marriott, S. & Leclair, S. (eds) *Fluvial Sedimentology VII*, Special Publication, International Association of Sedimentologists, 35, 61-74.

Gilvear, D., Winterbottom, S. & Sichingabula, H. (2000). Characters of channel planform change and meander development: Luanga River, Zambia. *Earth Surface Processes and Landforms*, 25, 421-436.

Gregory, K.J. (2006). The human role in changing river channels. *Geomorphology*, 79, 172-191.

Gurnell, A.M., Downward, S.R. & Jones, R. (1994). Channel planform change on the River Dee Meanders, 1876-1992. *Regulated Rivers: Research and Management*, 9, 187-204.

Gurnell, A.M. (1997). Channel change on the River Dee meanders, 1946 – 1992, from the analysis of air photographs. *Regulated Rivers: Research & Management*, 13, 13-26.

Ham, D. & Church, M. (2000). Bed material transport estimated from channel morphodynamics Chilliwack River, British Columbia. *Earth Surface Processes and Landforms*, 25, 1123-1142.

Hooke, J.M. & Redmond, C.E. (1992). Causes and nature of river planform change. *In*: Billi, P., Hey, R.D., Thorne, C.R. & Tacconi, P. (Eds.), *Dynamics of Gravel-bed Rivers*, Wiley, Chichester, 557-571.

Horizons Regional Council. (1999). Rangitikei River Scheme Asset Management Plan (Report No. 99/EXT/357).

Horizons Regional Council. (2000). Rangitikei River Scheme Review (No. 3) Final Report (Report No. 2000/EXT/397).

Horizons Regional Council. (2004). Water Allocation Project Rangitikei River (Report No. 2004/EXT/606).

Kondolf, G.M. (1997). Hungry water: effects of dams and gravel mining on river channels. *Environmental Management*, 21, 533-551.

Knox, J.C. (1987). Historical valley floor sedimentation in the Upper Mississippi valley. *Annals of the Association of American Geographers*, 77, 224-244.

Lane, S.N. & Richards, K.S. (1997). Linking river channel form and process: time, space and causality revisited. *Earth Surface Processes and Landforms*, 22, 249-260.

Lane, S., Chandler, J. & Richards, K. (1994). Developments in monitoring and terrain modelling small-scale river-bed topography. *Earth Surface Processes and Landforms* 19, 349–368.

Leopold, L.B. (1992). Sediment size that determines channel morphology. In: Billi, P., Hey, R.D., Thorne, C.R. & Tacconi, P. (Eds.), *Dynamics of Gravel-bed Rivers*. Wiley, Chichester, 297-311.

Leys, K.F. & Werritty, A. (1999). River channel planform change: software for historical analysis. *Geomorphology*, 29, 107-120.

Liébault, F. & Piégay, H. (2001). Assessment of channel change due to long-term bedload supply decrease, Roubion River, France. *Geomorphology*, 36, 167-186.

Liébault, F. & Piégay, H. (2002). Causes of 20th century channel narrowing in mountain and piedmont rivers of southeastern France. *Earth Surface Processes and Landforms*, 27, 225-444.

Liébault, F., Clement, P., Piégay, H., Rogers, C.F., Kondolf, G.M. & Landon N. (2002). Contemporary channel changes in the Eygues basin, southern French Prealps: the relationship of subbasin variability to watershed characteristics. *Geomorphology*, 45, 53-66.

McEwen, L.J. (1989). River channel changes in response to flooding in the upper River Dee catchment, Aberdeenshire, over the last 200 years. *In*: Beven, K., Carling, P. (Eds), *Floods: Hydrological, Sedimentological and Geomorphological implications*. Wiley, Chichester, 123-140.

Marston, R.A., Girel, J., Pautou, G., Piegay, H., Bravard, J. & Arneson, C. (1995). Channel metamorphosis, floodplain disturbance, and vegetation development: Ain River, France. *Geomorphology*, 13, 121-131.

Martin, C.W. & Johnson, W.C. (1987). Historical channel narrowing and riparian vegetation expansion in the Medicine Lodge River Basin, Kansas, 1871-1983. *Annals of the Association of American Geographers*, 77, 436-449.

Martin, Y. & Church, M. (1995). Bed-material transport estimated from channel surveys: Vedder River, British Columbia. *Earth Surface Processes and Landforms*, 20, 347-361.

NIWA (2010) http://wrenz.niwa.co.nz/webmodel/ accessed 20 September 2010.

Passmore, D.G., Macklin, M.G., Brewer, P.A., Lewin, J., Rumsby, B.T. & Newson, M.D. (1993). Variability of late Holocene braiding in Britain. *In*: Best, J.L. & Bristow, C.S. (Eds), *Braided Rivers*. Geological Society Special Publication 75, London, 205-229.

Palmer, M.A., Berhardt, E.S., Allan, J.D., Lake, P.S., Alexander, G., Brooks, S., Carr, J., Clayton, S., Dahm, C., Follstad, S.J., Galat, D.J., Gloss, S., Goodwin, P., Hart, D.H., Hassett, B., Jenkinson, R., Kondolf, G.M., Lave, R., Meyer, J.L., O'Donnell, T.K., Pagano, L., Srvastava, P. & Sudduth, E. (2005). Standards for ecologically successful river restoration. *Journal of Applied Ecology*, 42, 208-217.

Piégay, H., Darby, S.A., Mosselmann, E. & Surian, N. (2005). The erodible corridor concept: applicability and limitations for river management. *River Research and Applications*, 21, 773-789.

Richardson, J.M. & Fuller, I.C. (2010). Quantification of channel planform change on the lower Rangitikei River, New Zealand, 1949-2007: response to management? *GeoScience* 2010/2, 1-26. http://muir.massey.ac.nz/handle/10179/1618

Rinaldi, M. & Simon, A. (1998). Bed-level adjustments in the Arno River, central Italy. *Geomorphology*, 22, 57-71.

Rinaldi, M., Wyzga, B. & Surian, N. (2005). Sediment mining in alluvial channels: physical effects and management perspectives. River Research and Applications, 21, 805-828.

Roberts, C.R. (1989). Flood frequency and urban-induced channel change: some British examples. *In*: Beven, K.J. & Carling, P.A. (Eds), *Floods: Hydrological, Sedimentological and Geomorphological implications*. Wiley, Chichester, 57-82.

Rumsby, B.T. & Macklin, M.G. (1994). Channel and floodplain response to recent abrupt climate change: the Tyne basin, Northern England. *Earth Surface Processes and Landforms*, 19, 489-515.

Schwendel, A.C., Fuller, I.C. & Death, R.G. Assessing DEM quality for change detection and morphological budgeting in upland streams: experimentation towards a recommended model, *River Research & Applications* in press.

Shields, F.D., Simon, A. & Steffen, L.J. (2000). Reservoir effects on downstream river channel migration. *Environmental Conservation*, 27, 54-66.

Simon, A. & Rinaldi, M. (2006). Disturbance, stream incision and channel evolution: the roles of excess transport capacity and boundary materials in controlling channel response. *Geomorphology*, 79, 361-383.

Spink, A., Fryirs, K. & Brierley, G. (2009). The relationship between geomorphic river adjustment and management actions over the last 50 years in the upper Hunter Catchment, NSW, Australia. *River Research and Applications*, 25, 904-928.

Surfer (2002). *Manual*. Golden Software, Colorado

Wathen, S.J. & Hoey, T.B. (1998). Morphological controls on the downstream passage of a sediment wave in a gravel-bed stream. *Earth Surface Processes and Landforms*, 23, 715-730.

Werritty, A. & Ferguson, R.I. (1981). Pattern changes on a Scottish braided river over 1, 30, and 200 years. *In*: Cullingford, R.A. (Ed), *Timescales in Geomorphology*. Wiley, Chichester, 53-68.

Werritty, A. & Leys, K.F. (2001). The sensitivity of Scottish rivers and upland valley floors to recent environmental change. *Catena*, 42, 251-273.

Winterbottom, S.J. (2000). Medium and short-term channel planform changes on the Rivers Tay and Tummel, Scotland. *Geomorphology*, 34, 195-208.

Williamson, R.B. Smith, R.K. & Quinn, J.M (1992). Effects of riparian grazing and channelisation on streams in Southland, New Zealand. 1. Channel form and stability. *New Zealand Journal of Marine and Freshwater Research,* 26 241-258.

Wishart, D., Warburton, J. & Bracken, L. (2008). Gravel extraction and planform change in a wandering gravel-bed river: The River Wear, Northern England. *Geomorphology*, 94, 131-152.

In: River Channels: Types, Dynamics and Changes
Editor: Daniel A. Molina, pp. 165-173
ISBN 978-1-61324-148-6

Chapter 5

River Planform Analysis as a Tool for Neotectonic Investigation: Case Study from the Northern Pannonian Basin

Judit Petrovszki
Department of Geophysics and Space Reserch, Eotvos Lorand University, Budapest, Hungary

Abstract

Sinuosity calculations were made in the Upper Theiss region to draw conclusions on the neotectonic activity of the northern-north-eastern part of the Pannonian Basin.

The upstream part of the River Theiss, the Bodrog and its tributaries (Ondava, Laborc, Ung and Latorca), the Sajó and the Hernád flows in the study are. These rivers were meandering before the river controls. Their last, pre-regulated planforms were mapped in the second part of the 19th century, during the Second Military Survey of the Habsburg Empire.

The sinuosity values were computed along these rivers with different sample section size (window size). This window size changed from 5 km to 80 km. There is a given window size for each rivers, determined by the bank-full discharge, which connected to the neotectonics. Along these rivers, the characteristic window size was 15 or 20 km.

The places, where the river sinuosity changed, were compared to the structural lines of the "Atlas of the present-day geodynamics of the Pannonian Basin". Every place where the fault lines of the neotectonical map crossed the rivers, the sinuosities were changed, increased or decreased depending of the direction of the relative vertical movements.

Introduction

The study area is located in the north-eastern part of the Pannonian Basin, in eastern central Europe. Here flows the upper part of the Tisza (Theiss), the Sajo, the Hernád, the Bodrog and its tributaries, the Ondava, the Laborc, the Ung and the Latorca (Figure 1).

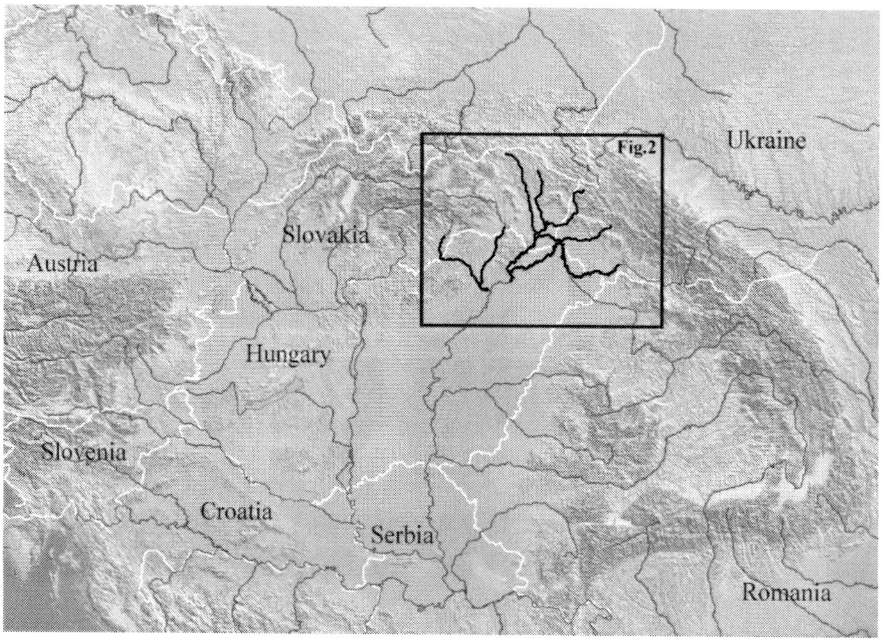

Figure 1. The study area is located in the north-eastern part of the Pannonian Basin.

The ancient river system was formed in Late Pliocene (Nádor et al., 2001, Gábris & Nádor, 2007). The subsidence rate changed often so the rivers often changed their channels. Now the subsiding is still working, but determined by the human activities. The main subsiding areas coincide with the main hydrocarbon and water extraction areas (Franyó, 1992, Joó, 1992).

The recent activity of the area were appointed by using the industrial seismic sections of the area, and presented on a neotectonic map (Horváth et al. 2006). But seismic sections are available only from the Hungarian part of the study area. And on the other hand, the upper part of the seismic sections is very noisy, because of the loose top layer (50-100 m thick). Therefore, it is hard to decide, whether the fault lines are really active or not. In these cases, the rivers can be very useful, because they can respond spectacularly to the changing elevation of the surface.

We can follow the past deformations by analysing the alluvial terraces, but it is not usable for recent tectonics. According to Schumm (1972, 1977, Schumm et al 1972), stream patterns are sensitive indicators of valley slope change. If the valley slope is decreasing, the sinuosity is also decreasing. Sinuosity is the ratio of channel length to valley length or valley slope to channel slope (Schumm, 1963). The range of sinuosity is from 1.0 (straight) to about 3.0. According to the flume experiments of Schumm and Khan (1972), five basic channel patterns exist, depending on the gradient, sediment load, stream power, and type of sediment load transported by the channel. These basic channel patterns are the straight, meandering channel of equal width, meandering channel that are wider at bends than in crossings, meandering-braided transition and typical braided stream. The range of channels from straight through braided forms a continuum, but the pattern changes occur at river pattern thresholds, at critical ranges of valley slope, stream power and sediment load (Schumm and Khan, 1972).

Ouchi (1985) was also made experimental studies to determine the effects of active tectonics on alluvial channels. Any tectonic deformation that changes the slope of a river valley results in a corresponding change in sinuosity to maintain the equilibrium channel slope. This effect has proved to be a diagnostic tool in identifying areas of active tectonics (Gomez and Marron, 1991). But channel pattern change alone is not sufficient evidence for active tectonics. Changes of hydrology, sediment load, human's activities and climatic variations also can cause channel pattern changes as well as active tectonics.

The theoretical and experimental results were tested on real rivers and creeks, since the 80's of the 20th century, mostly in the United States. Adams (1980) used it on the Mississippi and Missouri rivers. Burnett and Schumm (1983) applied this method in Louisiana and Mississippi to verify the uplift of the Monroe and Wiggins anticlines. The rivers of the coastal plain of Carolinas were analyzed by Marple and Talwani (1993, 2000). The connection between the rivers and the tectonics was demonstrated by Aswathi et al. (2008) in India, on the Pannagon River. In Slovakia, Pišut (2006) studied the Lower Morava River. In Hungary, several papers were written about the using of this method. Timár (2003) demonstrated the recent activity of the Mid-Hungarian Shear Zone (Csontos et al., 1992). Zámolyi et al. (2010) analyzed the river planforms of the Little Hungarian Plain, Petrovszki and Timár (2010) analyzed the Körös River System.

Nowadays the rivers in the Pannonian Basin are regulated, most of them during the second half of the 19th century. During these works the lengths of the rivers decreased and the channel slope increased. The shape of the rivers became straighter, and they are controlled, so they can not follow their own wish, they can not react to the valley slope changes. Because of the river controls, the map sheets of the Second Military Survey of the Habsburg Empire were used in this work. These sheets displayed the last pre-regulated planform of the studied rivers.

METHODS

Because of the river control works, in this paper the map sheets of the Second Military Survey of the Habsburg Empire (Hofstätter, 1989; Jankó, 2001; Timár, 2004; Timár et al., 2006) were used. These maps display the original channels of the rivers, because they were surveyed before the main river controls. But at some places, the planned cut-offs were also marked.

The Second Military Survey was made between 1806 and 1869. In the area of the former Hungarian Kingdom, it was measured between 1819 and 1869. They used Cassini projection, and the projection origin was at the Stephans Dom in Vienna. The geodetic datum is the Zach-Oriani hybrid ellipsoid. The scale of the map sheets is 1:28,800 (Hofstätter, 1989; Varga, 2002). The estimated accuracy of the map sheets to the modern system is 50-100 m, but most places, it remains under 30 m (Hofstätter, 1989; Jankó, 2001; Timár, 2004; Timár et al., 2006). The georeferring of the survey sections was made by using the database of Timár et al. (2006).

During the digitalization, a poly-line was draw along the middle of the river-beds, in the Hungarian National Grid (EOV) coordinate system. The digitized rivers were: Sajó (158 km), Hernád (111 km), Bodrog (134 km), Ondava (140 km), Laborc (115 km), Ung (107 km), Latorca (165 km) and Upper-Theiss (375 km). These river sections are altogether 1300 km long. The digitizing points were at different distances to each other. So before the sinuosity calculations, a conversion was needed, to get a series of point, with equal distance to their neighbours. This equal distance was 50 meter.

The river sinuosity values, the ratio of the channel length to the valley length, were calculated by the definition of Schumm (1963). Using an excel worksheet; the channel length was the "window size", changed from 5 km to 50 km, with a step of 5 km. The valley length was the Euclidean distance between the two endpoints of this "window". Using this method, the sinuosity values were calculated almost for all the points of the river. Except for the first few and last few kilometres of the digitized channels were not calculated, because of the calculation method. The sinuosity values related to the middle of the window, so at the two end of the channel for a half-window-size long section, the sinuosity could not computed.

All the sinuosity values, calculated by different window sizes were displayed on a spectrum-like diagram, "sinuosity spectrum" (van Balen et al., 2008). The horizontal axis represents the distance along the river; the vertical axis indicates the window size. The sinuosity values were illustrated using a gray-scale. If a sudden sinuosity change appears at several window sizes, it likely to be connected to the active tectonics.

A classification was also made on the data of the sinuosity values computed with different window sizes. Imported like a multichannel file, one window-size was one channel. On these data, an Unsupervised ISOCLASS Classification was made, using the ER Mapper program. During the classification, the sinuosity values were divided into 5 classes. Except the first and last 25 kilometres of the river, where there were not enough data for the classification, see above, for every point of the river a class was assigned. This method does not depend on the geometric parameters of the rivers.

If the river is too short, e.g. in this case less than 50 km, neither the classification, nor the sinuosity spectrum could be usable with these parameters. In this case, the window sizes should be ten times smaller. This window size should be change from 500 metres to 5 kilometres, with a step of 500 metres.

RESULTS

The sinuosity spectrums were created for all studied river sections (Figure 2).

Almost along the whole length of the River Ondava, the sinuosity values were medium (A1, A3, A5, A7, A9, A11). During some short section, the sinuosity was high (A6, A8) or low (A2, A4, A10). The River Laborc started with low sinuosity (B1), which increased gradually (B2). The section with high sinuosity values (B3) is parallel to the high sinuosity section of the Ondava (A8), and later, the low sinuosities B5 (along the River Laborc) and A10 (along the River Ondava), too. The River Ung (Uh) had low sinuosity along almost the whole digitized channel (C2, C4). Along two short sections (C1, C3), the sinuosity became medium. The significant sinuosity increasing appeared at C5, where the sinuosity values were medium and then high. The River Latorca also started with low sinuosity (D1). The sinuosity increased a bit (D2), but after decreased again (D3). The medium sinuosity values appeared in the section D4, and near the Hungarian-Slovakian-Ukrainian border the sinuosity was high (D5). Downstream the sinuosity became medium (D6) and high again (D7). The River Bodrog had medium sinuosity almost its whole lengths (E1 and E3). The sinuosity values were medium high in section E2, and low in section E4. The river Theiss had low sinuosity along the Ukrainian part of the digitized channel, and along the Hungarian-Ukrainian border (F1). Then the sinuosity was medium (F2) and low again (F3). In section F4, the sinuosity values were medium again, and then became high (F5). In the F6 section, the sinuosity decreased suddenly, and became low. In section F7 and F9, the sinuosity values were medium. Between them the sinuosity was low (F8). In the last part of the Upper-Theiss, the sinuosity values were medium low. The digitized part of the River Sajó started with low sinuosity values (G1). The section G2 had medium high sinuosity values. Downstream, the values decreased a bit (G3), but after, they increased again, and the river has medium sinuosity values (G4). In section G5 the sinuosity became low. Before the River Sajó met the River Hernád, the sinuosity increased again (G6). The River Hernád started with medium and medium low sinuosity (H1), and then, the sinuosity became low (H2). Before the Hernád flowed into the Sajó, the sinuosity of the river became medium high (H3).

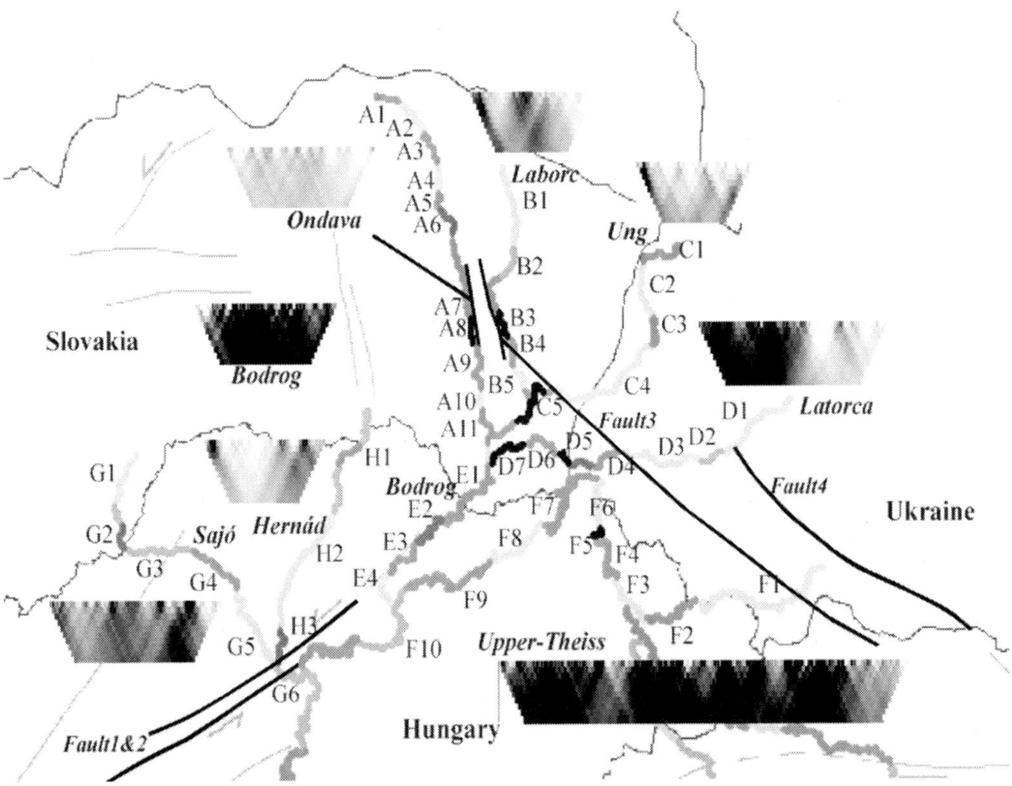

Figure 2. The studied rivers are displayed. The gray-scale indicates the sinuosity values. Darker gray means higher sinuosity values. The significant sinuosity changes are also displayed along the river channels. Light gray lines are the structural lines of the neotectonical map. Black lines indicate the recently active faults, verified by the river sinuosity changes.

Discussion

There is a window size related to the neotectonics, which is directly proportional to the bank-full discharge. If the river has higher bank-full discharge, this window size will be higher. In most cases, this neotectonic-window-size was about between 20 and 30 km. The main sinuosity changes of these neotectonic window-sizes were displayed on the river beds, using the same gray-scale (Figure 2). The results were correlated to the "Atlas of the present-day geodynamics of the Pannonian Basin" (Horváth et al, 2006) and some seismic sections. But seismic sections were available only in the Hungarian part of the study area. The results are in agreement with both the faults of the neotectonic map and the faults of the seismic section.

A seismic section, parallel to the River Sajó was studied, crossing the ***Fault1and 2***. These two fault line was also indicated on the seismic section. The northern block is uplifted with respect to the southern block. Upstream the fault, the river flow with low sinuosity values (G5). Downstream the fault, the sinuosity increased (G6), in agreement with the flume experiments of Ouchi (1985). The sinuosity would increase more, if the Sajó would not flow into the Tisza. The sinuosity increased right as the river crossed the faults. It is a clear indication that this change caused by the fault; therefore, it is still likely to be active.

Fault3 crosses the tributaries of the Bodrog. After they met the fault, their sinuosity increased (A8, B3, C5, D4,). According to the sinuosity values of the River Laborc, this fault line is not exactly at the right place. It should not cross the river between B3 and B4, but between B2 and B3.

Fault4 meets with the River Latorca, and increase its sinuosity a bit (D2). This is the end of the fault line, so the vertical movements are less than other places. So according to the river, this line is active recently.

After the rivers crossed these four structural lines, the sinuosity of the rivers was changed significantly. The reason of the changes was the vertical movement along the faults, like in the flume experiments of Ouchi (1985). At certain places, here along the Fault3, according to the rivers, the structural lines run a bit different ways like displayed on the neotectonical map. But the neotectonical activity of this fault line was verified by four rivers, the four digitized tributaries of the Bodrog. Every place, their sinuosity increased after they met the fault line. So the neotectonic activity of this fault was indicated by river sinuosity changes along hundred of kilometres.

At some places, where the sinuosity changes are significant, but no fault line indicated on the neotectonocal map, seismic sections should be analyzed to decide whether is there any fault line or not. The changing sediment load, and sediment type, (Leopold and Wolmann, 1957), the changes from bedrock controlled to alluvial or vice versa, or the change of the rate of channel slope and bankfull-discharge also could cause the changes in the sinuosity values.

During the classification, the results were almost the same (Figure 3). Except along the Fault1 and 2, where the classified part of the rivers are shorter, because of the calculation method, and did not meet the faults. Along the Faults 3 and 4, the rivers crossed the fault, and the points of the channels are in different classes on the two sides of the fault line. Here the river Ung is a bit short; it terminated before crossing the Fault3. But the other rivers are verified the neotectonic activity of this fault. If the classified parts of the rivers are long enough, the classification is as useable as the sinuosity spectrum. In addition, for the classification, we did not need the bankfull-discharge, and other geometrical parameters of the rivers.

CONCLUSION

There could be a connection between the river sinuosity changes and the neotectonic activity. This connection was studied and analyzed in this paper, on the rivers of the Upper-Theiss area. If a tectonical movement has vertical components, the rivers can respond to it. The rivers, by changing their shape and of course their sinuosity, try to keep their channel slope. Every place where the rivers crossed the faults of the neotectonic map, the sinuosity of the rivers changed, verifying the recent activity of these structural lines. At the places of the significant sinuosity changes, seismic section analysis is needed to decide, the changes were caused by fault lines or some hydrological changes (e.g. the increased the transported sediment).

Eight rivers were analyzed. Along the four fault lines indicated on the neotectonic map, in 6 points were detected significant sinuosity changes. These changes are verified the recent activity of the studied fault lines. According to the Ondava and Laborc, on of the faults (Fault3) is not running in the correct place on the neotectonic map. So the rivers can made correct the neotectonical maps.

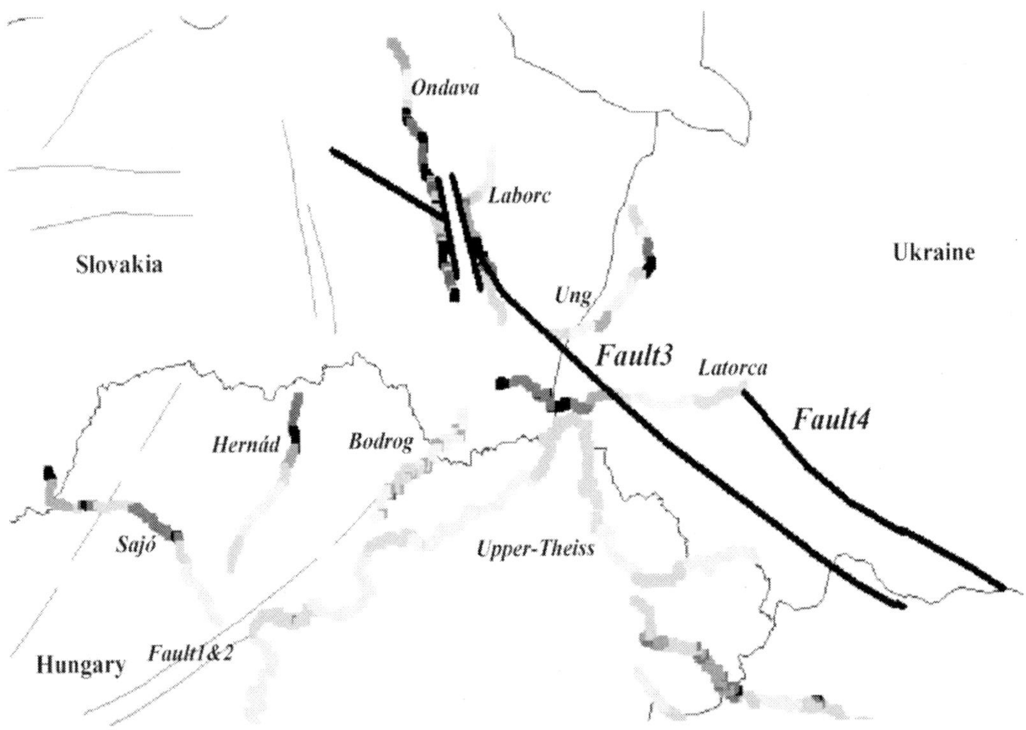

Figure 3. The results of the Isoclass classification are in agreement with the results of the sinuosity spectrum. The different gray colour along the river channels indicates the different classes.

The classification was also usefull, if the faults crossed the rivers along the analyzed section. In this case, the sinuosity spectrum was also computed, but the main parameters of the river did not need. We do not have to calculate a "neotectonic window-size" which depends on the bankfull-discharge. Only in case of small rivers and creeks, the window sizes should be ten times smaller.

REFERENCES

Adams (1980): Active tilting of the United States midcontinent: geodetic and geomorphic evidence. *Geology* 8: 442-446.

Aswathi, M. V., Vijith, H., Satheesh, R. (2008): Factors influencing the the sinuosity of Pannagon River, Kottayam, Kerala, India: an assessment using remote sensing and GIS. *Environmental Monitoring and Assessment* 138(1-3): 173-180.

Burnett, A. W., Schumm, S. A. (1983): Alluvial river-response to neotectonic deformation in Louisiana and Mississippi. *Science,* 222: 49-50.

Csontos, L., Nagymarosi, A., Horváth, F., Kovač, M. (1992): Tertiary evolution of the Intra-Carpathian area: a model. In: Ziegler, P. A. (ed.): Geodynamics of Rifting, Vol. 1. Case History Studies on Rifts: Europe and Asia. *Tectonophysics* 208: 221-241.

Franyó F. (1992): *Magyarország negyedidőszaki üledékeinek vastagságtérképe.* Térkép, méretarány=1:500.000, MÁFI, Bp.

Gábris, Gy., Nádor, A. (2007): Long-term fluvial archives in Hungary: response of the Danube and Tisza rivers to tectonic movements and climatic changes during the Quaternary: a review and new synthesis. *Quat. Sci. Rev.* 26, 2758-2782.

Gomez, B., Marron, D. C. (1991): Neotectonic effects on sinuosity and channel migration, Belle Fourche river, western South Dakota. *Earth Surface Processes and Landforms,* 16:227-235.

Hofstätter, E. (1989): Beiträge zur Geschichte der österreichischen Landesaufnahmen, I. Teil, Bundesamt für Eich- und Vermessungswesen, Wien, 196 p.

Horváth, F., Bada, G., Windhoffer, G., Csontos, L., Dombrádi, E., Dövényi, P., Fodor, L., Grenerczy, Gy., Síkhegyi, F., Szafián, P., Székely, B., Timár, G., Tóth, L., Tóth, T., (2006): A Pannon-medence jelenkori geodinamikájának atlasza: Euro-konform térképsorozat és magyarázó. *Magyar Geofozika* 47(4), 133-137.

Jankó, A. (2001): A második katonai felmérés. *Hadtörténeti Közlemények* 114: 103-129.

Joó, I. (1992): Recent vertical surface movements in the Carpathian Basin. *Tectonophysics* 202: 129-134.

Leopold, L. B., Wolman, M. G. (1957): River channel patterns; braided, meandering and straight. *USGS Prof. Paper* 282B: 1-73.

Marple, R. T., Talwani, P. (1993): Evidence of possible tectonic upwarping along the South Carolina coastal plain from an examination of river morphology and elevation data. *Geology,* 21: 651-654.

Marple, R. T., Talwani, P. (2000): Evidence for a buried fault system in the Coastal Plain of the Carolinas and Virginia – Implications for neotectonics in the southeastern United States. *Geol. Soc. Am. Bull.* 112: 200-220.

Nádor, A., Müller, P., Lantos, M., Tamóné, B. E., Kercsmár, Zs., Tóthné, M. Á., Sümegi, P., Farkasné, B. J., Nagy, T. (2001): Klímaváltozások és üledékes ciklusok kapcsolata a Körös-medence negyedidőszaki folyóvízi rétegsoraiban. *Földtani Közlöny* 130: 623-645.

Ouchi, S. (1985) Response of alluvial rivers to slow active tectonic movement. *Geol. Soc. Am. Bull.* 96: 504-515.

Petrovszki, J., Timár, G. (2010): Channel sinuosity of the Körös River system, Hungary/Romania, as possible indicator of the neotectonic activity. *Geomorphology,* 122, 223-230

Pišut, P. (2006): Evolution of meandering Lower Morava River (West Slovakia) during the first half of 20th century. *Geomorphologica Slovaca* 6(1): 55-68.

Schumm, S. A. (1963): *A tentative classification of alluvial river channels.* US Geol. Surv. Circ. V477.

Schumm, S. A. (ed., 1972): *River morphology.* Dowden, Hutchinson & Ross, Stroudsburg PA, 444 p.

Schumm, S. A. (1977): *The Fluvial System,* John Wiley & Sons, New York, 338 pp.

Schumm, S. A., Khan, H. R. (1972): Experimental study of channel patterns. *Geol. Soc. Am. Bull.* 83: 1755-1770.

Schumm, S. A., Khan, H. R., Winkley, B. R., Robbins, L. G. (1972): Variability of river patterns, *Nature* 237, 75-76.

Timár, G. (2003): Controls on channel sinuosity changes: a case study of the Tisza River, the Great Hungarian Plain. *Quat. Sci. Rev.* 22, 2199–2207.

Timár, G. (2004): GIS integration of the second military survey sections — a solution valid on the territory of Slovakia and Hungary. *Kartografické listy* 12, 119–126.

Timár, G., Molnár, G., Székely, B., Biszak, S., Varga, J., Jankó, A. (2006): *Digitized maps of the Habsburg Empire — the map sheets of the second military survey and their georeferenced version.* Arcanum, Budapest. 59 pp.

van Balen,R.T., Kasse, C.,Moor, J. (2008): Impact of groundwaterflowonmeandering: example from the Geul river, the Netherlands. *Earth Surf. Process. Landf.* 33 (13), 2010–2028.

Varga, J. (2002): *A vetületnélküli rendszerektől az UTM-ig. Kézirat, internetes elérhetőséggel.* (URL: http://www.tar.hu/vj1945bp/Osszes/Dok3uj.htm)Varga, 2002).

Zámolyi, A., Székely, B., Draganits, E., Timár, G. (2010): Neotectonic control on river sinuosity at the western margin of the Little Hungarian Plain. *Geomorph.* 122, 231–243.

In: River Channels: Types, Dynamics and Changes
Editor: Daniel A. Molina, pp. 175-193
ISBN 978-1-61324-148-6

Chapter 6

ASSESSMENT OF THE HYDRO-GEOLOGICAL SETTING OF THE NILE BASIN AND ITS IMPACT ON WATER RESOURCES MANAGEMENT

Mohammed El Bastawesy*, Safwat Gabr and Omar Cherif
Geological application division, National Authority for Remote Sensing and Space Sciences (NARSS), Cairo, Egypt

ABSTRACT

The Nile is the longest river on earth, it flows for nearly 6700 km from Latitude 4° S(the Equatorial Lakes) to Lat 32° north (the Mediterranean Sea) and it also drains the largest catchment in Africa, which equates an area of 31,1 million km^2. Proxy data of the fluvial sediments, cores of the Equatorial and Ethiopian lakes, age dating techniques, indicate that the river system has undergone great changes in recent geological history. The regional tectonic movements and geological evolution of the area have controlled the integration of different constituting catchments into the Nile mega drainage basin. Moreover, the Quaternary climatic changes have manifested drastic hydrological fluctuations, variability in the Nile flows and connectivity of different catchments via overflow. The Equatorial and Rift African lakes were desiccated during arid climatic phases, but extended several tens or even hundreds of metres above their present levels, and thus enlarged contributing catchment areas to the Nile system during more prominent Monsoon and wet phases. Because the downstream riparian of the Nile Basin (Egypt and Sudan) are mostly dependent on the Nile River, it is necessary to assess the impact of hydro-geological setting, climatic variability and water management strategies of the headwater regions in the Nile Basin. In this paper remote sensing data, digital elevation models (DEMs) were utilised to extract the catchment parameters, and to delineate storage areas for the Equatorial Lakes region. Available rainfall/discharge data are integrated with the delineated morphometrical parameters of the catchments to construct a hydrological model, which simulates the water balance of different interconnect basins. The total annual discharges of the basins are generally very low (less than 7 % of the total

---

* Contact: m.elbastawesy@narss.sci.eg, Geological application division, National Authority for Remote Sensing and Space Sciences (NARSS), 3 Joseph Bros Tito st., El Nozha El Gedida, P.O.Box 1564 Alf Maskan, Cairo, Egypt

annual rainfall). Although, the basin of the shallow Lake Kioga (5 m deep) is of negative hydrological contribution among the Equatorial Lakes region, where all the annual rainfall and considerable amount of bypass water overflown from Lake Victoria basin are usually lost by evaporation and evapotranspiration. The runoff coefficient and discharge can be significantly increased by draining the Kioga Lake and swamps through dredging and deepening certain overflow channel reaches. On the other hand, construction or towering of the regulation and hydropower dams on the equatorial lakes will be of adverse impact on the annual water discharge downstream. This includes the occasional reduction of flow required for filling up to designed storage capacities, and permanently increasing surface areas of water that is exposed to evaporation. Alternative sites across given stream rapids and falls were suggested as suitable for considerable dam-hydropower generations and minimal water storage.

INTRODUCTION

The hydrology of river basins has not only received intensive research, but also attracted widespread public, political, and economical concerns. The awareness is increasing for the interplay of hydrological processes and the society including; fluctuations of river flows, water management strategies and agriculture, scenarios of climate changes and the impact on water resources and the future of delicate deltaic systems, which are heavily populated worldwide. However initiatives, cooperation and treaties are globally recognized to manage and share water resources for the trans-boundary river basins. Most of the developed countries sharing a river basin, have not configured a common water master plan; this is driven by the diversity of hydrological processes, current and projected needs of water, and lack of data, cooperation, trust, etc. Overall understanding of the hydrological processes is necessary for effective water management in trans-boundary river basins. But this is a real challenge for the poor and developed countries which share flowing rivers. All of the aforementioned is typically the case of the Nile River basin, yet successful management of the Nile River upstream is critical for downstream countries particularly Egypt as the demands for water is increasing. The gap could be narrowed by the implementation of regional and common framework, which integrate multitude of research, techniques and adopted strategies. For example, the 10 riparian countries (Fig 1), which share the Nile River, are working toward a new agreement (the Nile Basin Initiative) which call for the fair and just utilization of the Nile water across the basin rather than on the basis of allocating water budgets, particularly for Egypt and Sudan. Therefore, it is of utmost importance to understand of Nile basin hydrogeology, of river flow characteristics, and of patterns of variability affecting flow volume, in order to disclose the water management strategies that can implemented to increase flow volumes and water/land uses efficiencies.

The Nile is the world longest rivers with a total length of 6670 km and drain a catchment area of 3.2 million km^2. The equatorial lakes plateau in Uganda is drained by tributaries forming the White Nile, which confluences with the Blue Nile (originating from the Ethiopian highlands) at Khartoum. Atbara River is another major tributary of the Nile system, which also has its headwater on the Ethiopian plateau and joins the main Nile course at Atbara to the north of Khartoum. Thereafter, the Nile flows through the Saharan Desert without any significant water supplies (Williams, 2009).

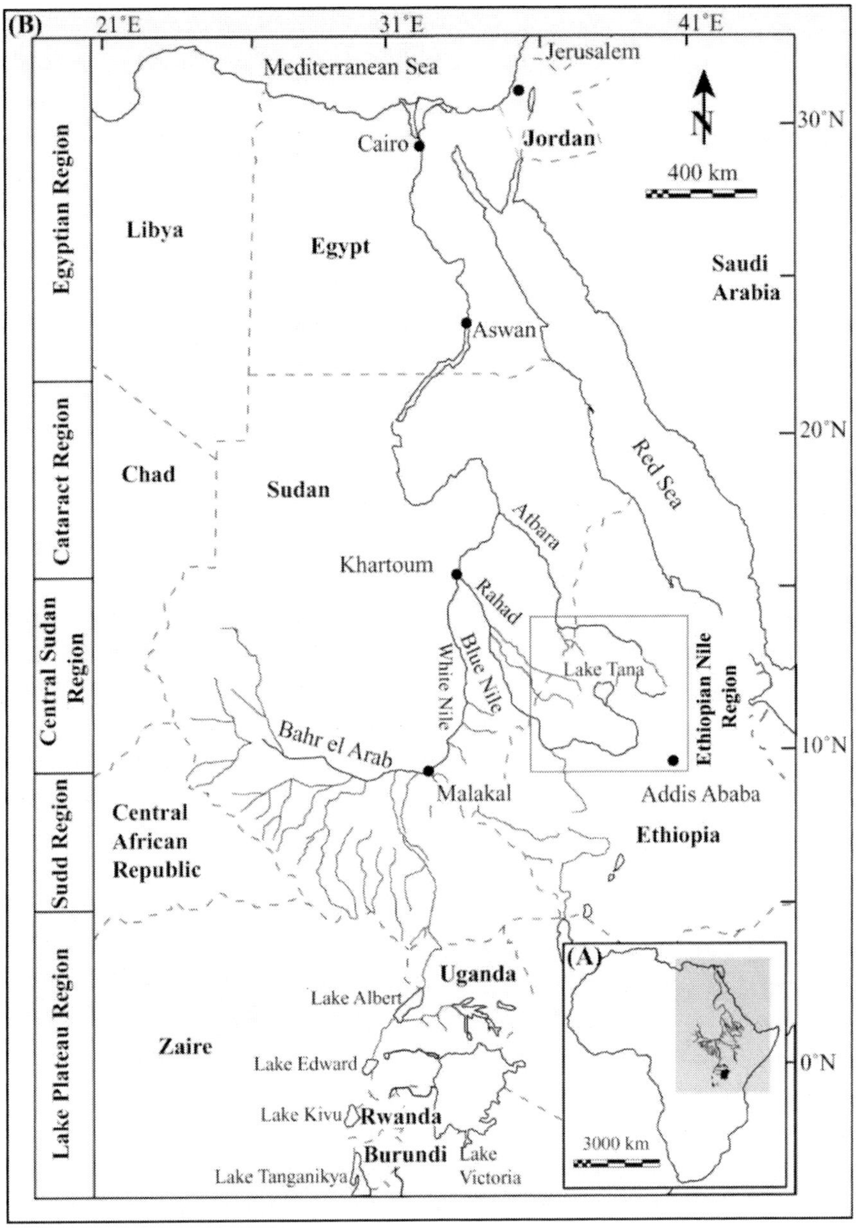

Figure 1. Location map of the Nile Basin.

Since the dawn of history, the inhabitants of arid lands in northern Sudan and Egypt mainly depended on the Nile water. It is also projected that by the year 2020 over 300 million people will sustain their livelihood in the Nile basin; therefore, a clear understanding of the Quaternary evolution of the basin, present hydrology, climatology and projected climatic changes scenarios are essential for the sustainable development plans (El Bastawesy et al., 2010).

The hydrological data of the Nile basin Cleary shows that amounts of rainfall, water seepage into groundwater, inflow and storage into the lakes, swamps and wetlands are largely high when compared with the river discharge downstream measured at Aswan High Dam. On average, a sum of 84 billion cubic meters of water is delivered to Lake Nasser upstream of the High Dam at Aswan, which was build in the 1969 to alleviate the disasters of high floods and low flows on the Nile Valley and the Delta of Egypt. The contribution of the different tributaries to the natural discharge at Aswan between the years 1912 and 1982 was estimated as follows; the Blue Nile (58%), the White Nile and the Subat (30%) and the Atbara River (12%) (Fig 2). During years of high discharge most of the additional water is contributed from the Blue Nile and the Atbara Rivers. For example, in the years of 1916 and 1946, which recorded exceptional high discharge records, these two rivers contributed 79 and 76 percent of the flow respectively (Said, 1993). On the other hand, the notable surge of rainfall only over the Lake Plateau during the 1960s was of limited impact on the flow discharge measured at Aswan. Most of the flows were retained in the source lakes and the Sudd of Sudan; the level of Lake Victoria has consequently risen by 1.9 meters.

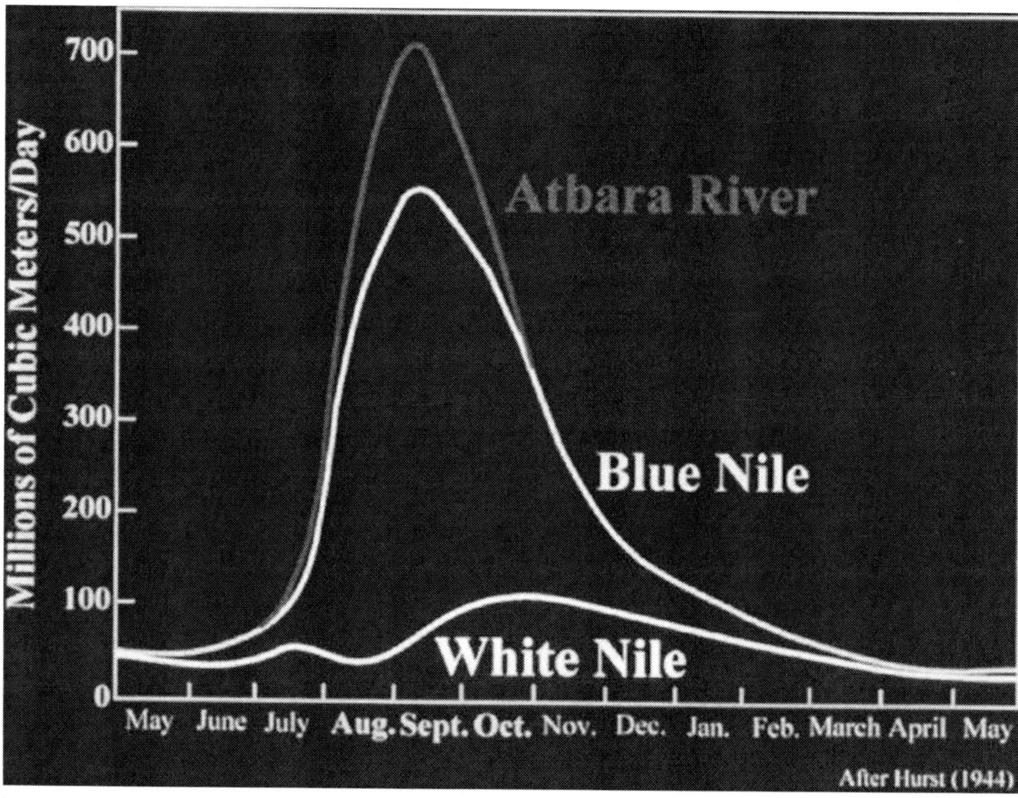

Figure 2. The annual average flow discharge of the main rivers of the Nile Basin at Aswan.

Overall, the fluctuation of the Nile flows is closely related to the Sahelian climatic events, which decrease during years of drought and increase during years of high rainfall. It is also noticed that these fluctuations may be related to the El Nino Southern Oscillation (ENSO) climatic events. Years of strong ENSO have been observed to be associated with high floods, severe storms and tidal surges in the Southern Hemisphere. Precipitation deficit usually occurs (from minus 5 to minus 15 percent) over the Nile Basin during El Nino years, while a more significant surge in rainfall anomalies from 10 to 25 percent, and from 5 to 10 percent over the Blue Nile and the White Nile respectively (Hulme, 1990, Janowiak, 1988).

Generally, in situ observations of the hydrological parameters over the entire basin are sparse and very limited. For example, measurements of rainfall are not available for large areas covered by lakes and swamps, which could be significantly different than that measured in the surrounding areas (e.g. Yin and Nicholson, 1998). However average monthly discharges at few key stations along the Nile, but the releases of several dams (particularly those recently constructed) are not published. Notwithstanding, data of evaporation and groundwater inflow are often not measured.

Fortunately, recent developments of Satellite remote sensing offer a complement to local measurements by providing observations that are temporally and spatially homogeneous. Therefore, averaged quantities can be estimated without the under-sampling problem inherent in sparse observation networks. Interestingly, multiple remote sensing data and products are free and publicly available (Swenson and Wahr 2009). Thus, several hydrological models were developed, to estimate the water balance of different sub-catchments of the Nile River. These various models were driven by the availability of remote sensing rainfall estimates, as such of the Tropical Rainfall Monitoring Mission (TRMM), monitoring of the lakes level fluctuation using satellite altimeter data such as Topex/Poseidon mission, and the Gravity Recovery and Climate Experiment (GRACE) satellite mission, which provide a monthly estimates of the earth gravity fields; these in turn can be used to estimate the total vertical integrated water storage(e.g. Awange et al., 2007; Birkett, 1995; Nicholson et al., 2003; Sutcliffe and Petersen, 2007; Swenson and Wahr, 2009).

Moreover, digital elevation models (DEM) of the Shuttle Radar Topography Mission (SRTM), have widely been used to extract different hydrographical parameters of the drainage basins; including drainage networks, catchment divides slopes gradient and aspect, and upstream flow contributing areas (e.g. Mark 1984, Tarboton, 1997).The available data archives of the Landsat MSS, TM and ETM+ starting from the year 1972 can be also used to monitor and assess the landuse and landcover changes.

THE AIM

The aim of this study is to assess the hydrogeological setting of the upstream sub-catchments of the Nile Basin and its impact on the water budget routing into downstream. There are several objectives to achieve this goal including1); to investigate the geological and paleo-hydrological development of the Nile Basin, particularly during the Quaternary. 2); to extract the different morphometrical parameters of the different sub catchments of the basin from the available SRTM DEM using GIS techniques. 3); to detect the changes of surface water bodies within the Nile basin using the analyses of available remote sensing images. 4); to assess the current climatological and hydrological data of the different sub-catchments, 5); to determine the geomorphological control on the water balance changes during routing from sources to outlet. 6); to assess the scenarios of storing and regulating flows of the Nile River on the hydrology and water budget downstream. 7); to suggest co-development strategies within particular areas to increase the yield of water budget and land use for the sustainable development of resources within the Nile Basin.

The Hydro-Geological Evolution of the Nile River

Multi-disciplinary research and investigations, systematic mapping, aerial photographs and recent satellite images had led to many findings and implications on the understanding of the geological and hydrological evolution of Nile Basin (Embabi, 2004). The current shape of the Nile River is a very recent development; it is inherited from the interconnection of several independent basins given the major climatic fluctuations that affected the entire globe, in addition to the tectonic movements, volcanic activities that affected different areas of the Nile Basin. It is postulated that the Nile in Egypt has excavated its main valley, which is similar to the modern Nile course during the Late Miocene, about 6 million years ago (Said, 1993). Thereafter, the Nile system has developed through several remarkable stages (i.e. phases), each was unique and characterized by a master river system, headwater extent, climatic conditions, tectonic activities, and sediment composition and supplies. These stages are termed (from old to recent), the Eonile, Paleonile , Protonile, Prenile, and Neonile, and these phases are described in more details in Said (1993) and with some modifications in Embabi (2004). It might be appropriate to give a short account on the different stages, although the early phases of the Nile ancestors had their headwaters carved in the Nubian massive of Egypt and north Sudan, without any connections with the African sources. Additionally, the given dates of the early stages are very approximate and relative and based on field investigations, morphological and sedimentological characteristics. But the recent dates (i.e. 50,000 years ago) are more accurate, as are mainly based on stable isotope contents of the primary lacustrine carbonates and organic matter, and biological remains (e.g., diatoms, ostracods, etc) of samples and deep-cores.

The Eonile

The Nile course in Egypt occupied a tectonic graben, which border the present valley of the Nile River, and was deeply entrenched during the Messinian time (Late Miocene) as a result of the desiccation of the Mediterranean Sea by several thousands of meters. Therefore, the Eonile (which was separated from the rest of Africa by the Nubian massif) formed a canyon larger and deeper than the Grand Canyon of the Colorado River; the deposits of this phase were only found in the subsurface of the delta embayment. They are composed of a lower clastic unit (sand and gravel) and upper unit of evaporites (Said, 1993). Concurrently, it is believed that considerable part of the Ethiopian plateau was drained into the Indian Ocean, though some flow entered the Sudd to the west. The Sudd was a giant basin, which formed an enormous lake having a depth of sediment up to 10 km thick. The Equatorial plateau was probably drained westward into the Congo basin, with some flow to the Indian Ocean or north into the Sudd (Sutcliffe and Parks, 1999).

The Paleonile

The Eonile Canyon was filled by water as the sea level rose during early Pliocene, the presence of this maritime estuary is inferred from the distribution of the Pliocene sediment in

Egypt. The lower marine facies (clays and silts) of the Paleonile attain a considerable thickness (1700 m), and they are overlain southward by fluviatile deposits. The extensive outcrops of the Pliocene flauvio-marine beds in Wadi El-natrun more than 100 km to the southwest of Cairo (i.e. apex of the present Nile delta) are thought to be of the Paleonile delta, which was five to eight times of the present continental delta. Furthermore, the distribution of the Paleonile sediments and the absence of central African fresh faunal elements in these sequences suggest that the Paleonile catchment was confined to local sources within the Red Sea Mountains of Egypt.

THE PROTONILE

However, the Paleonile flows were ceased by the prolonged arid climatic conditions that prevailed Egypt during the early Pleistocene. Minor and short wet periods (i.e. pluvial) and tectonic activities over the Red Sea area have re-organized drainage networks and developed a mighty river system termed 'the Protonile' (Butzer and Hansen, 1968; Issawi, 1978). The sediments of the Protonile are coarse (cobbles and gravel), and do not show any changes in their stratigraphy. Probably, the Protonile catchment extent was inherited from the preceding river system of the 'Paleonile'. The "Protonile" main river in Kom Ombo area had a course parallel to the present Nile at a distance 10-15 km to the west, and was at higher level of about 80-100 m above the present flood plain (El Bastawesy et al., 2010).

THE PRENILE

The short Protonile drainage pluvial was followed by a period of severe aridity during the Middle Pleistocene. But a fundamental reorganization of the drainage pattern occurred by the end of Middle Pleistocene; the Ethiopian connection was established and a new river (the "Prenile") was developed (Thurmond et al., 2004). Investigation of satellite images and available geological maps shows that tributary wadis in the Kom Ombo area display a complex pattern. Several discordant wadis are cross cutting each other, and also channels are frequently beheaded, offset and abandoned (Fig 3) (El Bastawesy et al., 2010).

These new drainage connections have continued to be the main source of water for the Nile up to the present time. Due to this outsource, huge quantities of sediments (Qena sands) were brought to Egypt, and covered large areas (stretches and patches) along the Nile Valley to the footslope of the Protonile gravels. The Prenile sediments are also present in both of the contemporary flood plain and the Delta at depths varying between 20 and 243 meters (Said, 1993). However some of these riverine sediments are also outcrop as low isolated yellow hillocks 'Turtle Backs' used to dot the cultivated fields of the Delta, but most of their sands were quarried. The mineralogy of these sediments differs from that of the earlier rivers, but resembles that of the present Ethiopian tributaries of the Nile (Shukri, 1950; Hassan, 1976). Indeed, the Prenile marks the first linkage of the local Egyptian drainage with its African tributaries (Embabi, 2004).

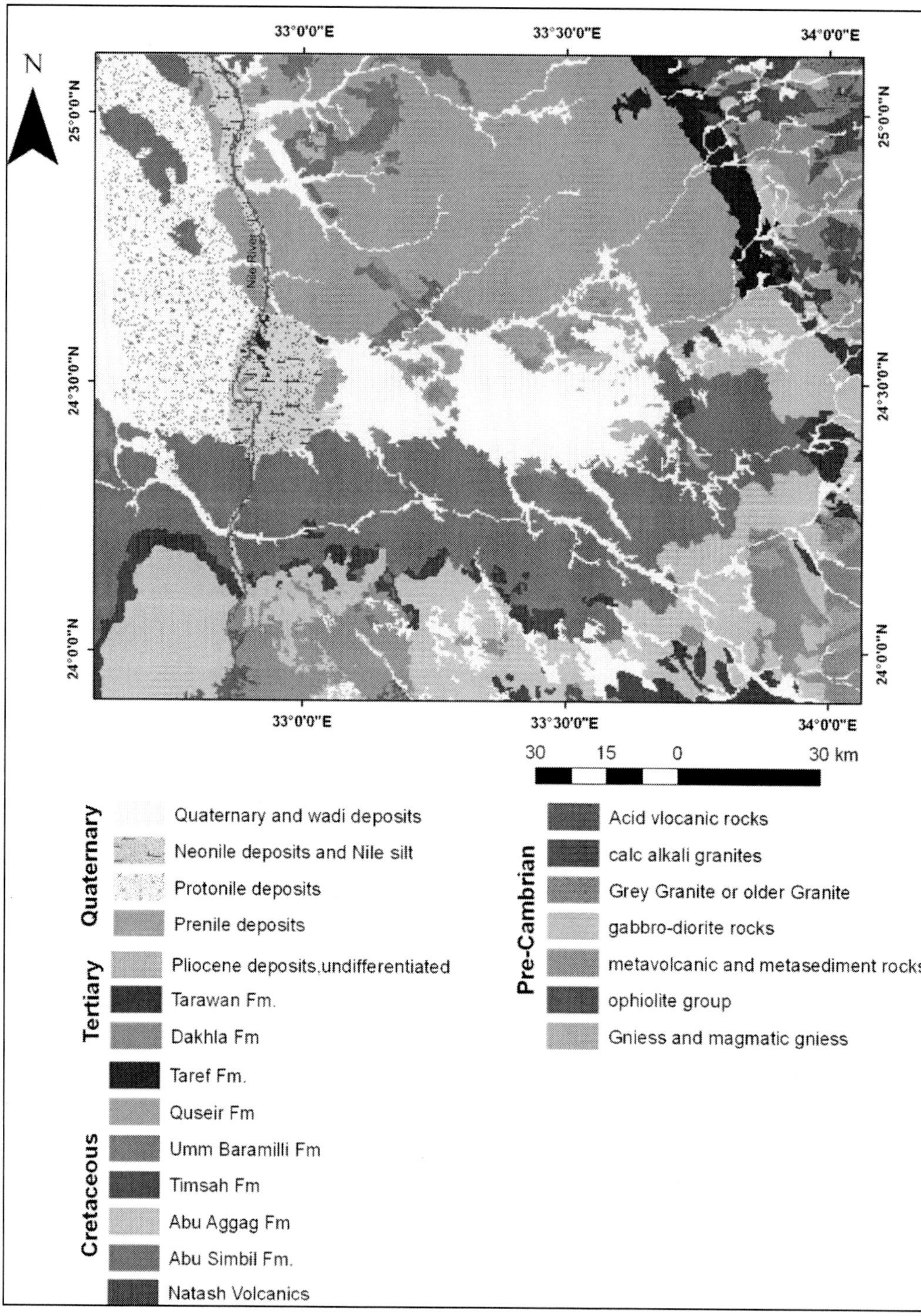

Figure 3. A geological map for the Kom Ombo area in Upper Egypt shows the complex and interlocked alluvium filled channels, which indicate tectonic re-organization of flow directions of the Nile tributaries in the different development phases.

THE NEONILE

The Prenile system was terminated due to the cutting of its African sources by the beginning of Late Paleolithic, and its sediments were locally capped by thick fluviatile gravel termed "Abbasiya gravel', which contain artifacts dated 400,000 years BP. The connection of the Neonile river with its African headwaters was tenuous and sporadic (Said, 1993). The flow extent of the Neonile was controlled by the climatic changes, which affected not only Egypt but also the Equatorial Africa, and the tectonic activities in the Nubian Swell, which continued to modify the gross hydrography of the Nile Basin well into the Pleistocene (Thurmond et al. 2004; El Bastawesy et al., 2010). Three events can be distinguished during the evolution of the Neonile. The earliest phase which probably started 400,000 years PB and continued until 200,000 years ago developed from a wet pluvial over Egypt. The African connection was severed and disrupted during a period of global cold climate (Mindel), (Said, 1993). But the connection with African sources was resumed again and deposited fine grained sediments 'the Dandara silts', which was totally different from the Prenile sediments. It may be important to mention that these silts are outcropping in the vicinity of the Nile floodplain in Upper Egypt, forming up to 23 meters high terraces, because of the down-cutting of the Nile River during consequent dry periods.

The second event of the Neonile (200,000 to 70,000 years ago) was associated with another wet pluvial, which also witnessed the appearance and spread of Middle Paleolithic man in Egypt. The local winter rains supplemented a low and erratic African Nile River. The third phase of the Neonile probably started 70,000 years ago to present, and it was associated with the last glacial period and the retreat of its ice sheets during the Holocene (some 10,000 years ago).

Interpretation of deep-core data and isotopes age dating techniques, showed that the Equatorial Lakes were desiccated During the last Glacial Maximum (from 20,000 to 15,000 years ago) (Talbot and Living-stone, 1989; Johnson et al., 1996; Gasse, 2000). The Equatorial forest had disappeared, the Sudd region was dry and the White Nile was probably obstructed and chocked with sand dune fields. Accordingly, the water level of Lake Victoria was at least 26 meters lower than it is today, and also the level of Lake Albert was at least 46 m lower than today (Beuning et al., 1997).

With the retreat of the ice sheets, the strong monsoon was returned and surged rainfall over the African headwaters of the Nile causing remarkable hydrological and landcover changes. For example the White Nile was probably occupied (to the south of Khartoum city) by a 400 km long and 20 to 30 km wide palaeo-lake, and it could has been maintained by significant increase in discharge produced from stronger monsoon 15000 years ago. This palaeo-lake was manifested by strandlines of palae-floods standing at several kilometres away from the present White Nile floodplain (Williams et al., 2006). Recent studies also suggested the Sudd area was formerly occupied by a massive lake during the Early Mid-Holocene (5,253-11,300 Ka P.B), probably at a high stand at the level of 515 m a.s.l. (El Shafie et al., 2011). Moreover, the contemporaneous closed lake of Trukana in the eastern branch of the Great African Rift Valley used to overflow into the White Nile sub-catchment (through the Sobat River) during the Mid-Holocene enhanced monsoon rainfall (Owens et al., 1982; Stanley et al., 2003; Williams et al., 2000).

From the aforementioned critical review for the hydrogeological development of the Nile Basin, it is obvious that the Nile catchment is a complex system, and composed of different basins, which were eventually interconnected and integrated into the modern system. There is now good evidence that the Nile River evolved to its contemporaneous shape as recent as 10,000 years ago. The entire catchment of the Nile equates an area of more than 3 million square kilometers and collects various regions, which are different from one another in relief, climate, geology, landuse and landcover. Therefore, scenarios of climate change, increasing demands for water resources particularly in the downstream dryland countries, and challenges to mandate a common framework of utilizing and developing the water of the Nile, all underpin the need to better implement the hydrogeological setting to determine management strategies that could suit and benefit the Nile partners.

Analyses of Remote Sensing and DEM Data

The hydrological significance of the Ethiopian and the Equatorial Lakes plateaus is well known as they provide the main source for the Nile flows. Herein a DEM covering the Equatorial Lakes was generated from the available SRTM data (Global Land Cover Facility, GLCF), and sets of multi-temporal satellite images of the TM and ETM+ were also downloaded (United Sates Geological Survey, USGS). Recent images of Egyptsat 1 for these areas were also obtained from the National Authority for Remote Sensing and Space Sciences (NARSS), Egypt.

Delineation of the Sub-Catchment and Drainage Networks

The SRTM data were first pre-processed, and the missing holes (i.e.gaps) were assigned the elevation values of the neighbouring pixles thorough a routine GIS spatial calculation, and then were complied into a mosaic which the cover the area of interest (Fig 4). The processing of DEM to produce hydrologically correct and connected drainage networks requires that sinks be first removed (Jenson and Dominique, 1988; Mark, 1984; Wise, 2000). This is can be achieved by filling all the sinks to the level of the next lowest cell of the perimeter. Filling a sink is an iterative technique. A given sink is filled to reach the nearest lowest elevation, the boundaries of the filled area may be a part of the new sinks, which then need to be filled and this process is repeated until all the pits are removed (ESRI, 1997). True sinks or closed depressions are thought and considered to be rare on natural surfaces and restricted to particular geomorphologic environments such as, those minor sinks developed on karst topography, or large closed lakes similar to those of the African Rift Valley. Following the standard procedures that treat sinks, even naturally existed, as artefacts in the DEM will seriously mislead the hydrographic delineation and consequent surface water budget analysis. Herein, the Equatorial Lakes are considered as natural sinks, as they are only hydrologically connected by overflows. Therefore, a few cells of the DEM were masked from these Lakes, and then the conventional automatic detection procedures of the D-8 method embedded in Esri software for the delineation of drainage networks and catchment parameters were typically followed (Fig 5). The steps are 1) filling sinks, then 2) flow direction, next 3), flow accumulation and finally 4) thresholding an up-slope contributing areas for the drainage channel and catchments (Jenson and Dominque, 1988). The obtained drainage networks and associated sub-catchments were overlaid ontop of the satellite images in order to be checked against visible wadis and streams. Furthermore, the mutli-temporal satellite images were visually interpreted to identify (if any) the recent landcover changes and the developed dams and reservoirs, in order to assess their potential impact on the Nile flows.

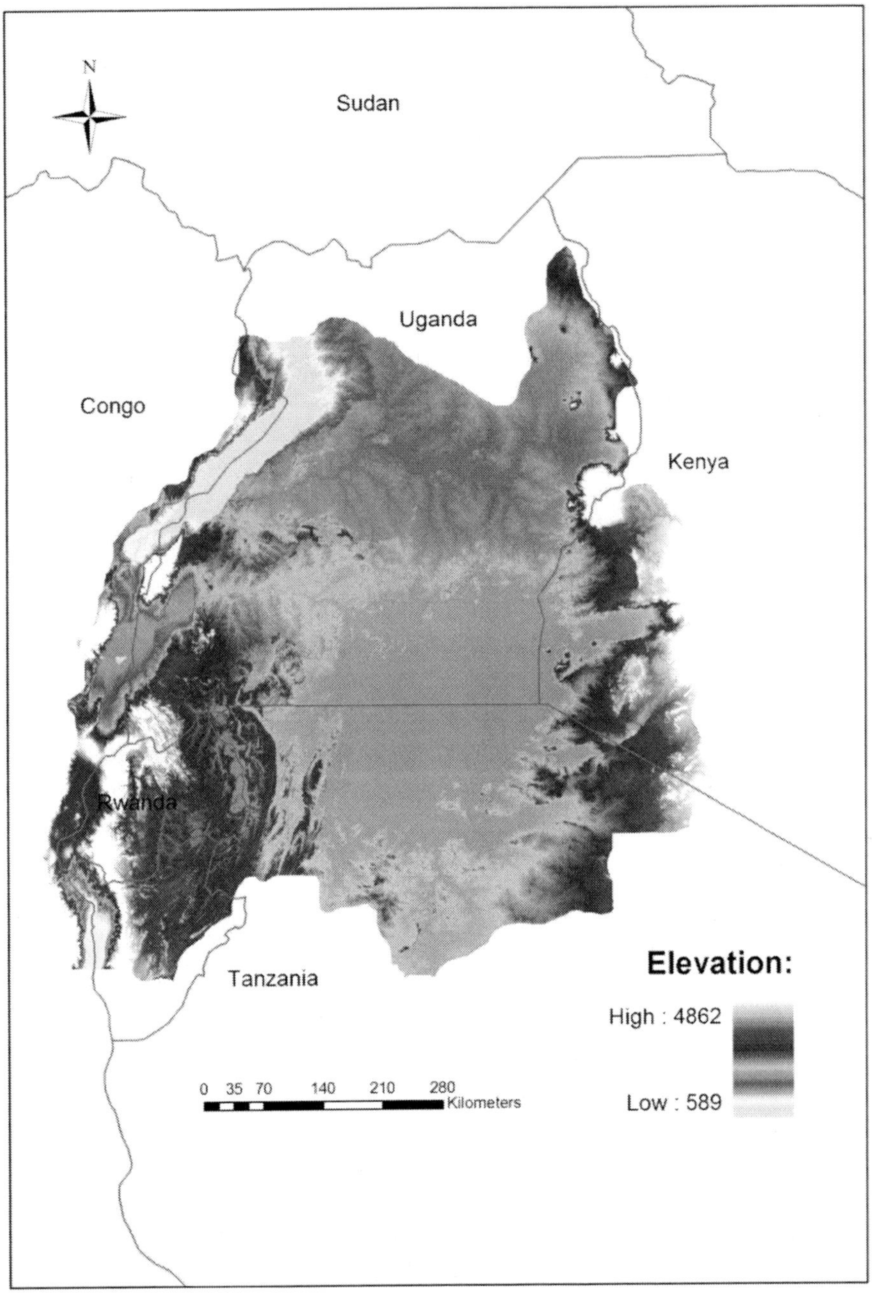

Figure 4. A processed SRTM DEM for the Equatorial Lakes region.

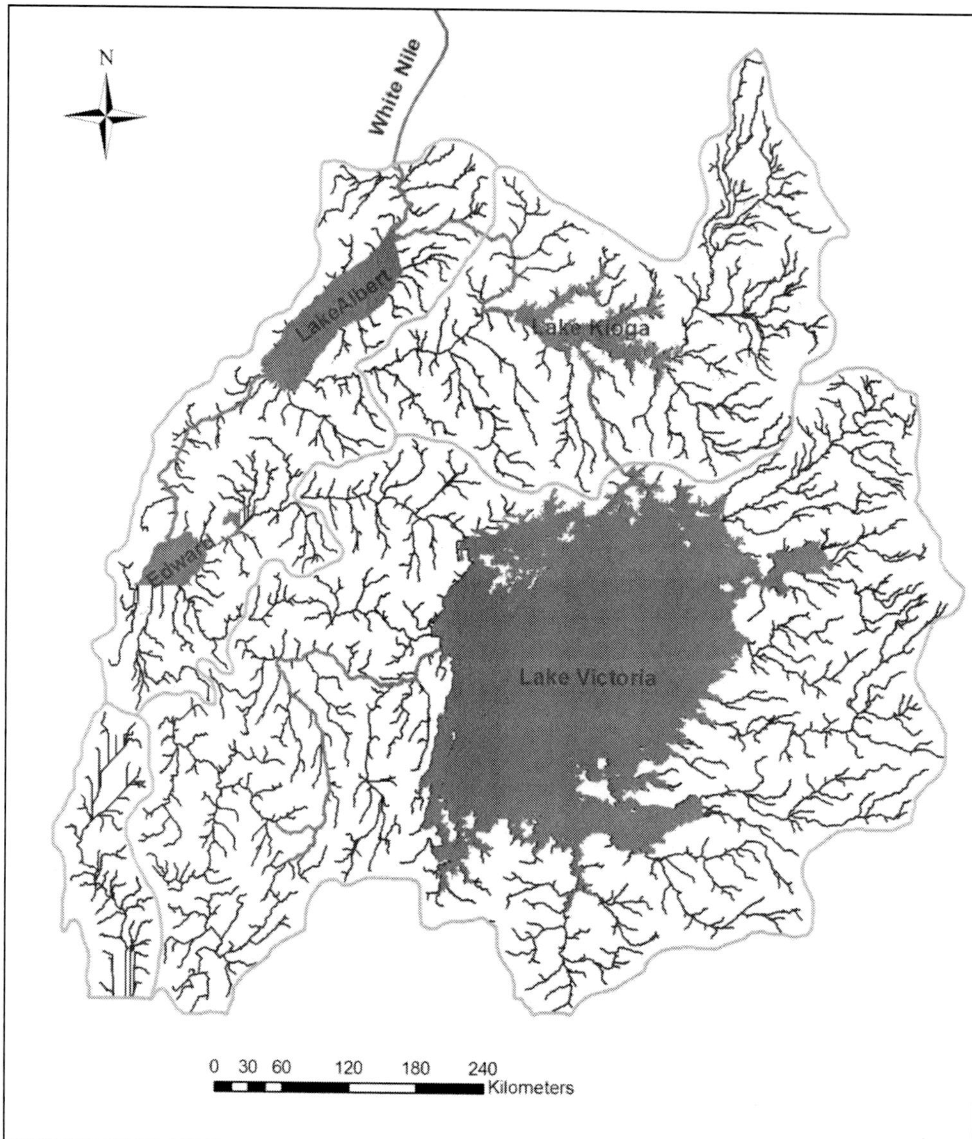

Figure 5. Shows the delineated lakes (blue areas) from satellite images, and the extracted drainage networks (blue lines) and delimited catchment boundaries (green lines) from the DEM analyses, note that these lakes are interconnected via the main river channels (red lines) that overflow the catchment divides in certain reaches.

The Water-Balance Flow Model for the White Nile System

The catchments of the Equatorial Lakes (Victoria, Kioga, and Albert) were delineated from the SRTM DEM and the lakes surface areas were delimited from the recent satellite images and the indicated hydrometrical data averaged from the available literature. Also for the sake of simplicity, the Lakes of Gorge and Edward (the smallest of the great African

Lakes), which drain into the Lake Albert via the Semliki River are included into the catchment of Lake Albert. Lake Victoria covers an area of 68,000 square km and ranked the second largest fresh water lake in the world. Despite of its enormous area; it's a shallow lake with a maximum depth of 80 m (Tate et al., 2001). The catchment area of the Lake Victoria is approximately 263,000 km^2. The estimated average rainfall over the catchment is 1150 mm per year, and the annual evaporation loss from the lake itself is approximately 1500 mm; the net annual and currently regulated overflow from the lake via Victoria Nile (at Ripon and Owen falls) into the swampy Lake Kioga is 23.5 billion cubic meters of water. However, this outflow is very sensitive to the variation of rainfall as evaporation rates do not vary greatly from year to year (Woodward et al., 2007) (Fig 6).

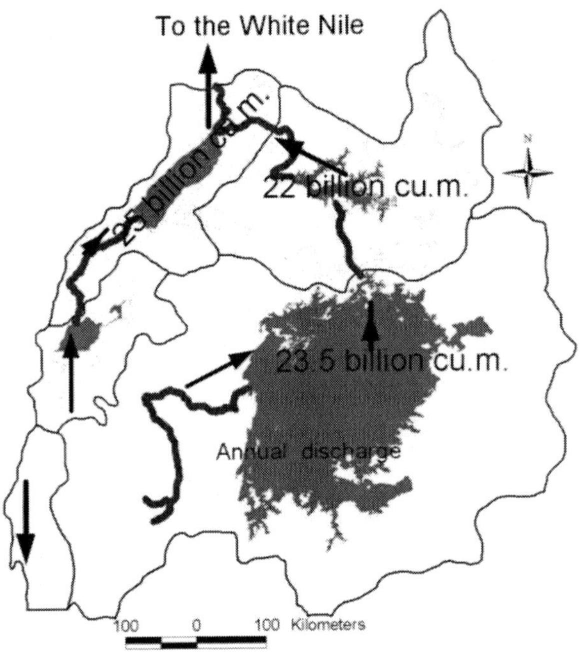

Figure 6. The annual discharge flow model from the Equatorial Lakes to the White Nile.

Lake Kioga is a large shallow swamp, which covers an area of 6,270 km^2, and the entire catchment equates and area of 75,000 km. the total average rainfall is estimated at 1500 mm per year, and the annual evaporation is 1290mm, while the net annual and unregulated overflow from the lake at Murchison falls into Lake Albert (including that received from Lake Victoria) is 22.5 billion cubic meters. Interestingly, the lake reaches a maximum depth of 7 meters, and most of the swamp is covered water lilies, and floating islands that drift between a numbers of small permanent islands. Furthermore the lake is a typical dendritic basin; it is represented by shallow drowned valley with its lower end (i.e. outlet) blocked by damming or tectonic titling (Loffler, 2004). Therefore, the excessive rains of 1997 and 1998 have resulted in exceptionally high water levels, which dislodged floating islands to accumulate at the lake's outlet. This blockage has developed a higher water level and flooded extra areas (580 km^2) of the swampy shorelines.

Lake Albert covers an area of 5,400 km^2 (about 160 km, long and 30 km wide), with a maximum depth of 51 m, and a surface elevation of 619 m above sea level. The delineated catchment of the Lake Albert basin equates an area of 37,000 km^2 and receives an estimated annual precipitation of 1100 mm, while the evaporation loss is approximately 1300 mm. the Lake Albert discharges about 26.5 billion cubic meters of water per years thorough the Albert Nile into the Sudd wetland of Sudan, where most of these water as well as input precipitation over the Sudd are lost by evaporation and transpiration (Fig 6). The outflows from the swamps are only half of the inflows, with very little seasonal variations (Sutcliffe and Parks, 1999).

The Proposed Schemes to Improve the Water Yield of the White Nile

However the White Nile contributes less than 30% of the total annual flows of the Nile River downstream at Lake Nasser, the hydrological potential of the basin is promising and can be developed. One of the early pioneer projects known as "the Jonglei Canal Project" was initiated to save some of the water loss into the Sudd swamps of the White Nile basin. The project idea was to construct a 280 km conveying channel (in the first phase) to bypass the swamp area between Jonglei and Malakal. The channel was set to save 4.7 billion cubic meters of water from the dispersion and loss into the Sudd. The digging of the canal began in June 1978, but unfortunately it came to a halt in 1983 as a result of the civil war in the region (Said, 1993). Different project ideas were also proposed to increase benefit of the artificially stored water bodies' thorough keeping minimal evaporation losses. For example, the Equatorial Lakes were proposed as alternative upstream reservoirs (where evaporation is minimal) instead of the artificial reservoirs downstream, which are characterized by high evaporation loss (Wittington et al., 2005). It was also urged that sedimentation problems within the Equatorial Lakes are much less pronounced than reservoirs on the Blue Nile or on the main Nile downstream (Guariso and Whittington, 1987). Among all the Equatorial Lakes, Lake Albert was singled out as most fit for storing water, as it lies at mostly contained within steep scarps of the rift system, and therefore its surface extent favours low evaporation losses per unit volume (Said, 1993). Indeed, multi-lateral efforts are needed to encourage and explore the different opportunities of cooperation and management of the Nile water. Therefore, this study proposes an additional project to improve the water discharge and optimal land resources management of the Equatorial Lakes area for the benefit of the riparian sharing the White Nile. In average a sum of 500 billion cubic meters of water is annually precipitated over this area, but only 26.5 billion cubic meters of water are discharged out of the basin into the Sudd region. Most of the rainfall is lost again by evaporation from the Equatorial Lakes and swamps, which cover more than 80,000 km^2. The low discharge of the Equatorial Lakes plateau is controlled by a unique hydrogeological setting where sets of different basins dotted by extensive lakes and swamps are only connected by overflows. This explains why the flows contributions of these lakes are more or less invariable with changes in the precipitation, unlike the Blue Nile flows which is highly sensitive to variations of rainfall. Also, the Quaternary climatic fluctuations have drastically affected the hydrological connectivity of these basins as well as the River Nile downstream. During the dry climatic cycles, the flow receded and the lakes were desiccated, but during the wet cycles the connectivity was resumed. This hydrological setting can be further exploited, as the swamps can be drained and the resulting flows can be stored in the lakes for hydropower generation and providing more discharge downstream. The catchment of Kioga (75,000 km^2) has a very high potential for this scheme. The huge annual rainfall (1500 mm) over the catchment, in addition to the inflow from Lake Victoria (23.5 billion cubic meters of water) are greatly affected by evaporation in the bypass swamp (5 meter deep) of Kioga, and only 22.5 billion cubic meters of water are discharged into Albert Lake. Having delineated the catchment boundary of each basin from the DEM analyses, certain stretches of the overflow channel courses (i.e. rivers) at the defined catchment divides can be mechanically dredged to increase discharge that will drawdown the swamp level. Draining a considerable part of the Kioga swamp will expand and improve agricultural productivity in vicinity as drainage will be enhanced; also the flow contribution of this huge negative hydrological catchment will be positive. These different projects require more deep, extensive and elaborate investigation and cooperation.

CONCLUSION AND RECOMMENDATIONS

In this research the hydrogeological evolution of the Nile Basin was reviewed and analyzed; to understand the impact of climatic fluctuations on the flows of the different sub-catchments of the river. The contemporaneous surface hydrological routines and are coupled to GIS and remote sensing that facilitate the interpretation and communication of results. The evolution of the Nile is complex and it was controlled by the tectonic movements related to the African Rift system and the Nubian massif during Tertiary and Quaternary times as well as the global climatic fluctuations. As a result, the upper reaches of the Nile River, particularly of the Equatorial Lakes were integrated and overflowed into the main river following the strong surge of Monsoon during the Holocene. Currently, the rivers of the Blue Nile and Atbara, which flow from the Ethiopian plateau, contribute with more than 70% of the total annual flow received downstream at Aswan. The flow discharge of these rivers is highly sensitive and correlated to the changes and fluctuations of rainfall over the plateau. On the other hand, the catchment of the White Nile is composed of sets of interconnected overflow-basins (Fig 7) which are occupied by massive lakes and swamps. However the White Nile catchment receive more than 500 billion cubic metes of rainfall water, it contribute very less flows downstream than the Ethiopian rivers. Moreover, most of this water is lost by evapo-transpiration from the extensive lakes and swamps in Uganda and Sudan. Various schemes were proposed to manage and control the White Nile flows, including storing and regulating flows of the Equatorial Lakes, and also projects to increase flows downstream thorough construing a bypass Jonglei channel in the Sudd region. The hydrogeological analyses of the Equatorial Lakes basin showed that the catchment of Kioga (75,000 km^2) receives more than 100 billion cubic meters of rainfall water that are totally lost by evapotranspiration from the shallow swamp of Kioga and surrounding canopies. Therefore, it is proposed here to increase the discharge and flow contribution of this catchment by the mechanical dredging and deepening a certain reach of the overflow channel (i.e. river) that connect Kioga swamp and Albert Lake. Draining of the Kioga swamp will significantly increase discharge downstream that can be utilized in the hydropower generation from small dams that can be constructed in the falls and rapids area on the course between

Kioga and Albert. Also, occasional flooding hazard of the low laying areas surrounding the swamp will be reduced, and the agricultural land productivity will substantially increase as drainage of the fields will be improved. Finally, the increase of runoff discharge from the upper Equatorial reaches will benefit the downstream countries. It is recommended to explore collaboration of the multidisciplinary research across the Nile Basin in order to define areas needed for sustainable developments in a way that will reflect positively to the co-riparian.

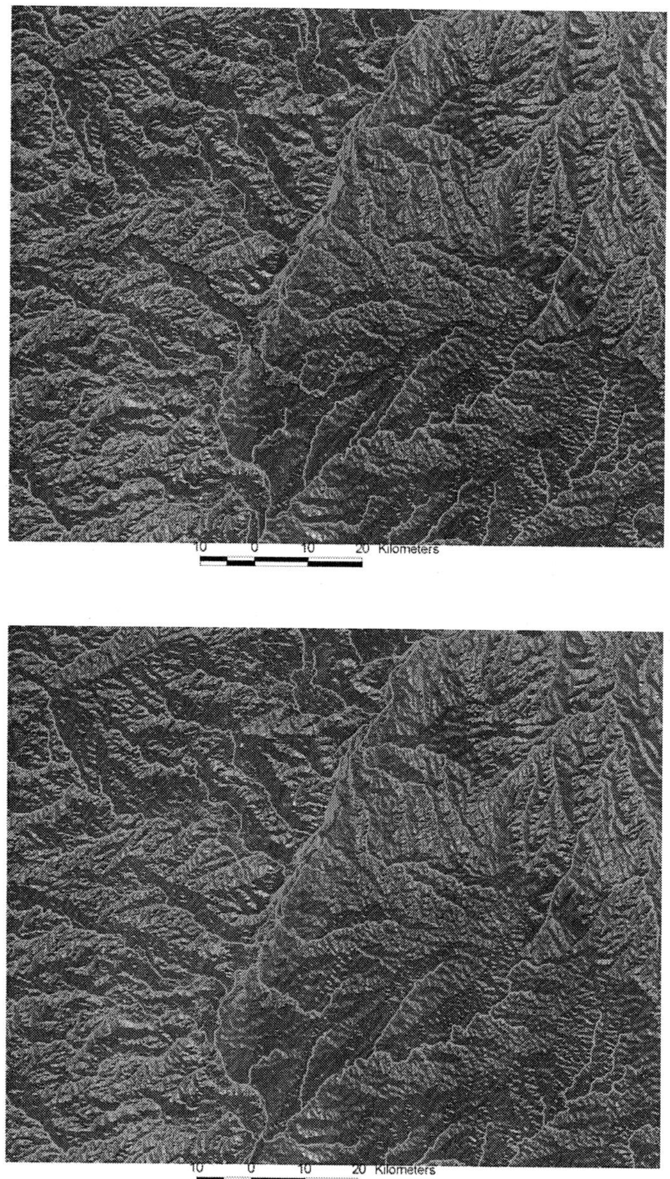

Figure 7. The extracted drainage networks (green) and water shed divides (blue) for the overflow reach between two different sub-basins in the Equatorial Lakes region. Top figure simulates the hydrographic networks of opposite flow directions among the watershed divides, and bottom figure demonstrate the overflow connectivity.

References

Awange, J.L., Sharifi, M., Ogonda, G., Wickert, J., Grafarend, E.W., Omulo, M., 2007. The falling Lake Victoria water levels: GRACE, TRIMM and CHAMP satellite analysis of the lake basin. *Water Resources Management*, 27, 775-796.

Beuning, K.R.M., Talbot, M.R., Kelts, K., 1997. A revised 30,000-year paleoclimatic and paleohydrologic history of Lake Albert, East Africa. *Palaeogeography, Palaeoclimatology, Palaeoecology*, 136, 259-279.

Birkett, C.M., 1995. The contribution of Topex/Poseidon to the global monitoring of climatically sensitive lakes. *Journal of Geophysical Research*, 100, 25179–25204.

Butzer, K.W., Hansen, C.L., 1968. *Desert and River in Nubia*. The University of Wisconsin Press: Madison; 562.

DWD, 2002. El Niño preparedness for Lake Kyoga and other flood prone areas of Uganda. *Directorate of Water Development. Ministry of Water, Lands and Environment,* Entebbe, Uganda.

El Bastawesy, M., Faid, A., and El Gammal, E., 2010. The Quaternary development of the tributary channels to the Nile River at Kom Ombo area, Eastern Desert of Egypt, and their implication on groundwater resources. *Hydrological Processes*, 24, 1856-1865.

El Shafie, A. A.I., Zeinelabdein, K., A. E., and Eisawi, A.A.M., 2011. Paleogeographic evolution and paleoenvironmental reconstruction of the Sudd area during the Early-Mid Holocene, Sudan. *Journal of African Earth Sciences* (in press).

Embabi NS. 2004. The Geomorphology of Egypt, Landforms and Evolution: Volume 1 (The Nile Valley and The Western Desert). The Egyptian Geographical Society: Cairo; 447.

ESRI, 1997. Watershed delineator application - User's Manual, Environmental Systems Research Institute, Redlands, CA.

Gasse, F., 2000. Hydrological changes in the African tropics since the Last Glacial Maximum. *Quaternary Science Reviews*, 19, 189-211.

Guariso, G., and Whittington, D., 1987. Implications of Ethiopian water development for Egypt and Sudan. *International Journal of Water Resources Development*, 3, 105-114.

Hassan, F. A., 1976. Heavy minerals and the evolution of the modern Nile. *Quaternary Research*, 6, 425-444.

Hulme, M., 1990. Global climate change and the Nile Basin. In: Eds Howell, P.P., and Allan, J.A., *The Nile, Resources Evolution, Resource Management, Hydropolitics and Legal Issues*, School of Oriental and African Studies (SOAS) and Royal Geographical Society, London, 59-82.

Issawi, B., 1978. Quaternary geology of Bir Tarfawi, Western Desert, Egypt. *Annals of the Geological Survey of Egypt*, 8, 295–304.

Janowiak, J.E., 1988. An investigation of international rainfall variability in Africa. *Journal of Climate*, 1, 240-255.

Jenson, S.K., and Domingue, J.O., 1988. Extracting topographic structure from digital elevation data for geographical information system analysis. *Photogrammetric Engineering and Remote Sensing*, 54,1593- 1600.

Johnson, T.C., Scholz, C.A., Talbot, M.R., Kelts, K., Ricketts, R.D., Ngobi, G., Beuning, K., Ssemmanda, I., McGill, J.W., 1996. Late Pleistocene desiccation of Lake Victoria and rapid evolution of Cichlid fishes. *Science*, 273, 1091-1093.

Loffler, H., 2004.The Origin of Lake Basins. In Eds: O' Sullivan, P.E., and Reynolds, C.S., *The Lakes Handbook: Limnology and Limnetic Ecology, Volume 1,* Blackwell Publishing Ltd.

Mark, D.M., 1984. Automatic detection of drainage networks from digital elevation models. *Cartographica*, 21, 168-178.

Nicholson, S.E., Some, B., Mccollum, J., Nelkin, E., Klotter, D., Berte, Y., Diallo, B.B., Gaye, I., Kpabeba, G., Ndiaye, O., Noukpozounkou, J.N., Tanu, M.M., Thiam, A., Toure, A.A., Traore, A.K., 2003. Validation of TRMM and other rainfall estimates with a high-density gauge dataset for West Africa. Part II: validation of TRMM rainfall products. *Journal of Applied Meteorology*, 42, 1355–1368.

Owens, R.B., Barthelme, J.W., Renaut, R.W., & Vincens, A., 1982. Palaeolimnology and archaeology of Holocene deposits northeast of Lake Turkana, Kenya. *Nature*, 298, 523–529.

Said R., 1993. *The Nile River: Geology, Hydrology and Utilisation. Pergamon Press*: Oxford, 320.

Shukri, N. M., 1950. The Mineralogy of some Nile sediments. *Quarterly Journal Geological Society of London*, 105, 511-534.

Stanley, J. D., Krom, M.D., Cliff, R.A., and Woodward, J.C., 2003. Short Contribution: Nile Flow Failure at the End of the Old Kingdom, Egypt: Strontium Isotopic and Petrologic Evidence. *Geoarchaeology: An International Journal*, 18, 395-402.

Sutcliffe, J.V., Parks, Y.P., 1999. *The Hydrology of the Nile*, IAHS Special Publication No. 5. IAHS Press, Institute of Hydrology,Wallingford, Oxfordshire.

Sutcliffe, J., Petersen, G., 2007. Lake Victoria: derivation of a corrected natural water level series. *Hydrological Sciences Journal*, 52, 1316–1321.

Swenson, S., and Wahr, J., 2009. Monitoring the water balance of Lake Victoria, East Africa, from space. *Journal of Hydrology*, 370, 163-176.

Talbot, M.R., Livingstone, D.A., 1989. Hydrogen index and carbon isotopes of lacustrine organic matter as lake-level indicators. *Palaeogeography, Palaeoclimatology, Palaeoecology*, 80, 283-300.

Tarboton, D.G., 1997. A New method for the determination of flow directions and contributing areas in grid Digital Elevation Models. *Water Resources Research*, 33, 309–319.

Tate, E.L., Sene, K.J. and Sutcliffe, J.V., 2001. Water balances study of the upper White Nile basin flows in the late nineteenth century. *Hydrological Sciences Journal*, 46, 301–318.

Thurmond AK, Stern RJ, Abdelsalam MG, Nielsen KC, Abdeen MM, Hinz E. 2004. The Nubian Swell. *Journal of African Earth Sciences*, 39, 401–407.

Williams, M., Talbot, M., Aharon, P., Abdl Salaam, Y.,Williams, F., Inge Brendeland, K., 2006. Abrupt return of the summer monsoon 15 000 years ago: new supporting evidence from the lower White Nile valley and Lake Albert. *Quaternary Science Reviews*, 25, 2651-2665.

Williams, M.A.J., 2009. Late Pleistocene and Holocene environments in the Nile basin. *Global and Planetary Change*, 69, 1–15.

Williams M.A.J., Adamson, D.A., Cock, B., McEvedy, R., 2000. Late Quaternary environments in the White Nile region, Sudan. *Global and Planetary Change*, 26, 305–316.

Wise, S.M., 2000. Assessing the quality for hydrological applications of digital elevation models derived from contours. *Hydrological Processes*, 14, 1909-29.

Wittington, D., Wu., X., and Cadoff, C., 2005. Water resources management in the Nile Basin: the economic value of cooperation. *Water Policy*, 7, 227-252.

Woodward JC, Macklin MG, Krom MD, Williams MAJ. 2007. The Nile: evolution, quaternary river environments and material fluxes. In *Large Rivers: Geomorphology and Management* , Gupta A (ed). John Wiley and Sons: Chichester; 263–292.

Yin, X., Nicholson, S.E., 1998. The water balance of Lake Victoria. *Hydrological Sciences Journal*, 43, 789-811.

INDEX

#

A

B

C

D

E

F

G

H

I

Q

R

S